Advances in Industrial Control

T0137286

Magdalene Marinaki and Markos Papageorgiou

Optimal Real-time Control of Sewer Networks

With 65 Figures

 Springer

Magdalene Marinaki, PhD
Markos Papageorgiou, PhD

Department of Production Engineering and Management,
Dynamic Systems and Simulation Laboratory,
Technical University of Crete, Chania, Greece

British Library Cataloguing in Publication Data
Marinaki, Magdalene, 1970–
    Optimal real-time control of sewer networks. — (Advances in
    industrial control)
    1. Combined sewers — Automatic control   2. Combined sewers —
    Automatic control — Mathematical models — Case studies
    3. Real-time control   4. Combined sewer overflows —
    Mathematical models — Case studies   5. Non-linear control
    theory
    I. Title   II. Papageorgiou, M. (Markos), 1953–
    628.2'14

Marinaki, Magdalene, 1970–
    Optimal real-time control of sewer networks / Magdalene Marinaki, Markos Papageorgiou.
        p. cm. — (Advances in industrial control)
    Includes bibliographical references and index.
    ISBN 1-4471-5673-0 (alk. paper)
        1. Combined sewers—Automatic control.   2. Real-time control.   I. Papageorgiou, M.
    (Markos), 1953–   II. Title.   III. Series.

TD662.M37   2004
628'.214—dc22   2004056451

Advances in Industrial Control series ISSN 1430-9491
ISBN 1-4471-5673-0 Springer London Berlin Heidelberg
Springer Science+Business Media
springeronline.com

MATLAB® and Simulink® are the registered trademarks of The MathWorks, Inc., 3 Apple Hill Drive,
Natick, MA 01760-2098, U.S.A. http://www.mathworks.com

The use of registered names, trademarks, etc. in this publication does not imply, even in the absence
of a specific statement, that such names are exempt from the relevant laws and regulations and therefore
free for general use.

The publisher makes no representation, express or implied, with regard to the accuracy of the information contained in this book and cannot accept any legal responsibility or liability for any errors or
omissions that may be made.

Typesetting: Electronic text files prepared by author

69/3830-543210 Printed on acid-free paper SPIN 10952702

Advances in Industrial Control

Professor Emeritus O.P. Malik
Department of Electrical and Computer Engineering
University of Calgary
2500, University Drive, NW
Calgary
Alberta
T2N 1N4
Canada

Professor K.-F. Man
Electronic Engineering Department
City University of Hong Kong
Tat Chee Avenue
Kowloon
Hong Kong

Professor G. Olsson
Department of Industrial Electrical Engineering and Automation
Lund Institute of Technology
Box 118
S-221 00 Lund
Sweden

Professor A. Ray
Pennsylvania State University
Department of Mechanical Engineering
0329 Reber Building
University Park
PA 16802
USA

Professor D.E. Seborg
Chemical Engineering
3335 Engineering II
University of California Santa Barbara
Santa Barbara
CA 93106
USA

Doctor I. Yamamoto
Technical Headquarters
Nagasaki Research & Development Center
Mitsubishi Heavy Industries Ltd
5-717-1, Fukahori-Machi
Nagasaki 851-0392
Japan

To my brother, to my father, and to the memory of my mother

Magdalene Marinaki

To Elia

Markos Papageorgiou

Series Editors' Foreword

The series *Advances in Industrial Control* aims to report and encourage technology transfer in control engineering. The rapid development of control technology has an impact on all areas of the control discipline. New theory, new controllers, actuators, sensors, new industrial processes, computer methods, new applications, new philosophies . . . , new challenges. Much of this development work resides in industrial reports, feasibility study papers and the reports of advanced collaborative projects. The series offers an opportunity for researchers to present an extended exposition of such new work in all aspects of industrial control for wider and rapid dissemination.

Water is a valuable resource in all societies and its provision is only the first part of the water cycle. The second part involves collecting wastewater and then treating it so that its subsequent discharge to the environment is not harmful in any way. This collection is accomplished by sewer networks and the treatment is usually biological, taking place in dedicated wastewater treatment plants. These systems usually have a fixed maximal capacity and the problem is that the system is subject to uncontrolled, unpredictable large-scale disturbances in the form of rainfall. Sudden surges of influent swelled by storm waters can have a devastating effect on these systems, since if the surge does not bypass the wastewater treatment plant then all the biologically active media therein can be flushed into the environment. It can take weeks to restore these plants to biologically effective levels. The problem is to find cost-effective ways of dealing with these periodic surges of inflow. One remedy is the use of storm-water-storage basins but a highly pertinent question is whether the storage capability and the control of the sewer network itself can be exploited as an alternative or supplementary remedy. Magdalene Marinaki and Markos Papageorgiou (Technical University of Crete) set out to investigate this question and the result is this excellent monograph that breaks new ground in this control applications field. Whilst the regulatory framework for wastewater treatment has rapidly become more stringent, the technology to deliver the new regulatory demands has emerged more slowly. Control engineering has a significant contribution to make to this field as this monograph shows.

The monograph opens by introducing the reader to the models and simulation tools for sewer networks. In Chapter 3, a clear multi-level control strategy is proposed for sewer network flow control and this is followed by full development of the proposed approach. The creation of an optimization framework for sewer network control is detailed and solved in Chapters 4 and 5. The monograph is notable for the presentation of an extended case study based on the Obere Iller (Bavaria, Germany) sewer network. A key conclusion is that this approach can manage storm water flows quite successfully and a more inclusive global approach to sewer network and wastewater treatment plant control deserves further investigation.

The monograph will be of interest to all working in the water and wastewater treatment industry. It is an excellent example of how control engineering can help to solve some of the problems of this industry. Control engineering readers will find the application and the solution procedures proposed by Dr Marinaki and Professor Papageorgiou of considerable interest. The volume is an exemplary entry in the *Advances in Industrial Control* monograph series.

M.J. Grimble and M.A. Johnson
Industrial Control Centre
Glasgow, Scotland, U.K.

Preface

During recent decades an increased interest in the protection of the environment from everything that could lead to its downgrading and destruction has been observed. The overflows from combined sewer networks are clearly one of the main pollutant sources in the environment. The development of control systems minimising overflows of combined sewer networks aims at the protection of the quality of waters that receive the outflows of the networks.

This monograph gives a detailed description of the development, application, and simulation testing of an advanced control system for central sewer network flow control. A multilayer control structure that consists of three control layers (adaptation, optimization, and direct control) may be used for the control of a combined sewer network. With regard to the optimization layer, several approaches have been proposed in the past. This monograph is focused on the development and comparison of two methods for the optimization layer, namely the *nonlinear optimal control* and the *multivariable feedback control* methods. An important feature of this monograph is that the efficiency of the control methods used is investigated for a large-scale combined sewer network located at the river Obere Iller in Bavaria (Germany) through simulation with a realistic model using different scenarios of external inflows. This study was the basis for the implementation of these control strategies in the particular sewer network.

This book is aimed at control and water engineers; researchers in the fields of control, modelling, and simulation of combined sewer networks; scientists who are involved in the design, development, and implementation of control systems for combined sewer networks; and postgraduate students working in the field of sewer network modelling and control.

Magdalene Marinaki and Markos Papageorgiou
June 2004

Acknowledgements

The authors would like to express their thanks to Dr.-Eng. Albert Messmer for valuable discussions and for making available the simulation program KANSIM. The support of Mr. P. Schaad and of the Obere Iller Verband are also acknowledged.

Finally, the authors would like to thank Prof. Michael A. Johnson (University of Strathclyde, Glasgow) for the encouragement to proceed with the preparation of this monograph.

Table of Contents

Chapter 1
Introduction

The pollution (the surcharge of materials or energy or microorganisms that are pathogenic for humans and animals) of the groundwater and underwater is one of the most important problems that preoccupies people and authorities around the world. There are many ecological consequences of the pollution of the groundwater. For example, the physicochemical characteristics of the water may be changed, leading, among others, to serious economical consequences, *e.g.*, an increase in the cost of water processing for its reuse.

The most important problems of pollution concern the water (lakes, rivers, and the sea) that suffers the strongest exploitation and use. One of these uses is as receivers of the outflows of combined sewer networks (Seidl *et al.*, 1998; Lee and Bang, 2000; Chebbo *et al.*, 2001). The construction of treatment plants, to enable sewage treatment before disposal, protects the quality of the water that receives the outflows of the networks. However, urban combined sewer networks do not have separated collectors for the domestic and industrial sewage, on the one hand, and the rainwater drainage, on the other hand. Therefore, during rainfall, networks or treatment plants may be overloaded, and overflows may take place upstream of overloaded stretches, causing pollution of receiving waters. Placing retention reservoirs at appropriate locations of the network [by constructing special basins (off-line storage) or by installing throttle gates at the end of voluminous sewer stretches (in-line storage)] is a cost-efficient way to avoid overflows in moderate rain events and to reduce them in stronger rainfall, as the water is stored in the reservoirs during the rainfall and is directed toward the treatment plant after the end of the rainfall.

Optimal operation of a combined sewer network (that contains retention reservoirs) (Figure 1.1) implies that for each rain event the whole retention capacity of all reservoirs will be used before overflows take place somewhere in the network. However, this cannot be guaranteed by fixed gate settings, such as fixed weirs or manually adjustable gates for the filling and emptying of the storage spaces. Especially if the rainfall is distributed unevenly over the urban area,

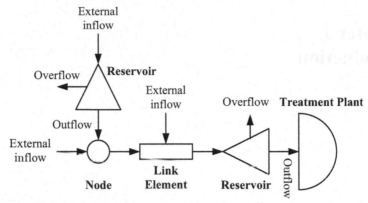

Figure 1.1. Schematic representation of a small sewer network.

there may be reservoirs that are not totally filled, while overflows already occur elsewhere in the network. In these cases, a further considerable reduction of overflows can be obtained by real-time operation of the reservoirs, *e.g.*, by use of controllable gates. The decision on how to move the gates during a certain rain event may be taken by a human operator or by an automatic control strategy to be applied in real time. An efficient control strategy can reduce substantially the overflows from a sewer network. In addition, it may lead to substantial cost savings as the number and storage capacities of the reservoirs required to keep overflows below a certain limit [usually legislatively defined (Zabel *et al.*, 2001)] depends on the efficiency of the applied control strategy.

A structure for real-time control of sewer networks that combines high efficiency and low implementation cost can be composed of a number of control layers (multilayer control structure). Such a *flexible* hierarchical structure, because of its modular character, that is *reliable* due to its decentralized structure and *efficient* due to the real-time operation and the global consideration of the overall network, provides real-time control of sewer networks. It consists of three control layers, as has been proposed by Papageorgiou and Mevius (1985): An *adaptation layer* is responsible for inflow and possibly rain prediction (if needed) and for real-time estimation of the system state. An *optimization layer* is responsible for the central, overall network control (*i.e.*, for specifying reference trajectories for the reservoir storages and outflows). A decentralized *direct control layer* is responsible for the realization of the reference trajectories. With regard to the optimization layer, several approaches have been proposed in the past, like nonlinear optimal control, multivariable feedback control, methods based on dynamic or linear programming, expert systems, fuzzy control, and further heuristic approaches.

Nonlinear optimal control is potentially the most efficient approach due to direct consideration of inflow predictions, of process nonlinearities, and of constraints. On the other hand, nonlinear optimal control implies development and implementation of sophisticated codes for real-time numerical solution of the optimal control problem. *Multivariable regulators*, if designed properly, may approximate the efficiency of nonlinear optimal control, based on much simpler calculation codes. The approaches that are based on *dynamic programming* are difficult to apply to large-scale networks due to the "curse of dimensionality", while the approaches based on *linear programming* may not include the nonlinearities of the process. *Expert systems, fuzzy control*, and further *heuristic approaches* are less efficient, less systematic, and more cumbersome to develop, to maintain, and to expand than theoretically founded approaches.

All the previously mentioned approaches have some special characteristics that call for thorough study and analysis before they can be implemented in the optimization layer. The need for developing and applying an effective central control method was the motivation for carrying out the research presented in this monograph, which describes the development, application, and simulation testing of a control system for central sewer network flow control. More precisely the following issues are addressed in some detail:

- The development and analysis of two general and systematic methods for the central control of sewer networks, namely a nonlinear optimal control and a multivariable feedback control method. Several improvements, modifications, and extensions are introduced to previously developed versions of the methods in order to increase their efficiency.
- Application of both control approaches to a particular combined sewer network located at the river Obere Iller in Bavaria (Germany), which connects five neighboring cities to one treatment plant. The efficiency of the control methods is investigated for this network through simulation with a realistic model using different scenarios of external inflows, and a comparison of their respective results is undertaken.

The structure of this book is as follows. Chapter 2 presents the mathematical modelling of combined sewer networks. Chapter 3 presents the objectives of combined sewer network control and provides an extended literature review of the methods proposed for the particular control problem in the past as well as the approach pursued in this research. Chapters 4 and 5 present the nonlinear optimal control and the linear multivariable feedback control, respectively, as a basis for central sewer network flow control. Chapter 6 describes the application of the developed methodologies to the particular real sewer network of Obere Iller in Germany. Chapter 7 presents the results obtained for the specific sewer network by use of both approaches, and compares them, as well as the results obtained with the no-control case. Finally, Chapter 8 summarizes the main conclusions.

Chapter 2
Modelling of Sewer Network Flow

2.1 Introduction

The first step in the study and control of a process is the development of a mathematical model of the process behaviour. The *mathematical model* includes a set of equations that describe, with more or less accuracy, the behaviour of the process, and through this model (using appropriate input data) the time evolution of internal quantities of the process may be calculated. The mathematical model of the process may be derived via:

- The *deductive way*, by using known laws of physics that describe the relevant aspects of the studied process, as for example the continuity equation. In the case of a sewer network, the model that is deduced in this way is referred to as the *hydraulic* or *hydrodynamic model* (Béron and Richard, 1982).
- The *inductive way*, by using experimental results (input–output values) that indirectly characterize the behaviour of the process.
- A combination of both previously mentioned approaches, that is, the use of both physical laws and experimental results. A mathematical structure derived from physical laws may be fitted to the data, as it is the case for the sewer network flow when *hydrological models* (Béron and Richard, 1982) are developed.

Frequently, for the same process it is necessary to derive more than one model with different degrees of accuracy and complexity. Typically two models are developed in control applications. The first is a sufficiently accurate simulation model that describes the behaviour of the process with high realism in order to be used for testing the performance of the control methods. The second model has lower accuracy and complexity and is used for the design of the controller, leading to accordingly limited design complexity and moderate computational effort.

In this book, two mathematical models are employed for the study of the sewer network control problem: a realistic simulation model that is referred to in the sequel of the monograph as the *accurate model of the sewer network*, and a simpler control design model that is referred to as the *simplified model of the sewer network*.

Both employed models address only the water flow dynamics in a wastewater system, not the water quality aspects and dynamics. This is in accordance with the addressed control problem, which concerns water quantities, not water quality dynamics (see Chapter 8 for suggested future extensions). It should be noted that most commercially available modern software packages for the simulation of wastewater systems describe not only the water flow but also the pollutant dynamics in the system. The development of real-time control systems, however, that explicitly consider water quality measurements and aspects are still in their infancy (Ashley *et al.*, 1999).

2.2 Accurate Model of Sewer Networks

Combined sewer networks consist of a set of elements in which different processes take place, as for example storage (in the reservoirs or in the sewers), transportation (in the sewers), and merging of flows (in the nodes). All these processes in the different elements of the sewer networks are modelled using known laws of hydraulics for the development of the accurate model of the sewer network.

In the following subsections, a few typical elements are described, upon which most combined sewer networks may be built.

2.2.1 Link Elements

a) Hydrodynamic Link Elements

The hydrodynamic link element (the hydrodynamically modelled sewer), is used where a nonnegligible storage of volume is induced in a sewer stretch by spillback or by flow regulation via throttle gates.

The basis for the mathematical modelling of the hydrodynamic link element are the Saint-Venant equations (Mays and Tung, 1978), namely the continuity equation and the momentum equation, which describe with satisfactory accuracy the dynamic behaviour of the flow in the sewers of the network. The Saint-Venant equations take into account the friction and the inertia of the flowing liquid and can also describe backwater phenomena (Labadie *et al.*, 1980). The flow process is regarded as one-dimensional current whereby the dependent variables are the flow q, the velocity of flow v, and the flow depth h, while the independent variables are the distance x and the time t (Figure 2.1). The following continuity equation, expressing the conservation of mass in the flowing liquid, corresponds to the first Saint-Venant equation:

$$\frac{\partial F(h)}{\partial h}\frac{\partial h(x,t)}{\partial t}+\frac{\partial q(x,t)}{\partial x}=0 \tag{2.1}$$

where:
- $q(x,t)$ is the flow (in m³/s) at location x (in m) along the sewer axis at time t (in s).
- $h(x,t)$ is the depth of sewer flow (in m) at location x (in m) along the sewer axis at time t (in s).
- $F(h)$ is the cross-sectional area of flow (in m²) (Figure 2.2).

The second Saint-Venant equation, the momentum equation, with which the resistance to flow is expressed, is derived from the equation $\upsilon = q/F(h)$ for flow velocity $\upsilon(x,t)$ (in m/s) and Bernoulli's theorem for instationary currents (Duncan *et al.*, 1970)

$$\frac{1}{g}\frac{\partial \upsilon}{\partial t}+\frac{\upsilon}{g}\frac{\partial \upsilon}{\partial x}+\frac{\partial h}{\partial x}=I_S-I_R \tag{2.2}$$

where:
- I_S is the sewer slope.
- I_R is the friction slope or the slope of the line of energy of flow [see (2.3) below].
- g is the gravitational acceleration (in m/s²).

In the momentum equation (2.2) we have the following terms:
- $\partial h/\partial x$ is the slope of the water surface.
- $(\upsilon/g)\cdot(\partial \upsilon/\partial x)$ expresses the influence of the change of the energy height $\upsilon^2/2g$ along the sewer (when the flow is rapidly changing in space).
- $(1/g)\cdot(\partial \upsilon/\partial t)$ expresses the influence of the time-change of the velocity (when the flow is locally changing rapidly over time, the flow is instationary).

I_R is calculated empirically from the equation (Manning's formula) (Labadie *et al.*, 1980)

$$I_R = cq^2 R(h)^{-4/3} F(h)^{-2} \tag{2.3}$$

where c is a constant (in s²/m^{2/3}) that depends on the sewer's characteristics and is equal to $(n/1.486)^2$, where n is Manning's roughness coefficient, and R(h) is the hydraulic radius of flow (in m) and is equal to $R(h) = F(h)/P(h)$, where P(h) is the wetted perimeter (in m) (Figure 2.2).

The first two terms in the momentum Equation (2.2) can be neglected in most practical cases as they are considered negligible in comparison to the third term (Wanka and Königer, 1984), in which case we have

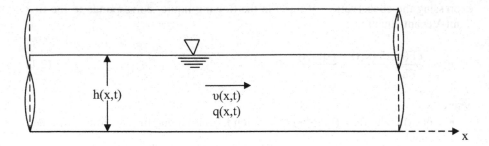

Figure 2.1. Variables in one-dimensional flow.

$$\frac{\partial h}{\partial x} = I_S - I_R.$$ (2.4)

Using (2.3) and (2.4), the following equation for the flow is derived

$$q = kR(h)^{2/3} F(h) \sqrt{I_S - \frac{\partial h}{\partial x}}$$ (2.5)

where the constant k is equal to $1/\sqrt{c}$. The Equations (2.1) and (2.5) are the basis for the hydrodynamically modelled sewer.

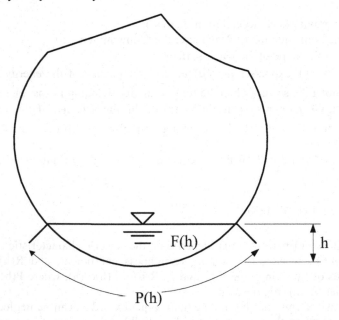

Figure 2.2. Cross section of a sewer with arbitrary shape.

b) Time-Lag Links

The hydrodynamic link elements are used for modelling relatively flat sewers where backwater phenomena may occur. However, for the mathematical modelling of sewers with relatively steep slope, namely $I_S>0.001$, and insignificant backpressure, a time-lag link model can be used. In this case, the slope of the water surface in the sewer is neglected $(\partial h/\partial x \approx 0)$ (uniform flow) and Equation (2.5) becomes an algebraic function:

$$q(h) = kR(h)^{2/3} F(h)\sqrt{I_S}. \tag{2.6}$$

Based on (2.6), the initial continuity Equation (2.1) is written, after appropriate transformations,

$$\frac{\partial h}{\partial t} + c(h)\frac{\partial h}{\partial x} = 0 \tag{2.7}$$

where

$$c(h) = \frac{\partial q(h)}{\partial h} / \frac{\partial F(h)}{\partial h}. \tag{2.8}$$

The partial differential Equation (2.7) has the following solution:

$$h(x,t) = h(0, t - \frac{x}{c}). \tag{2.9}$$

This means that the state at point x and time t equals the state that existed at point 0 at time $t - x/c$. This corresponds to state transfer (kinematic wave) with velocity c(h) that is given from Equation (2.8) and it may be assumed constant with respect to the flow conditions when q(h) is nearly linear in a range of water level values (Papageorgiou and Messmer, 1985). The time lag x/c applies in the same manner to the flow q(x,t) as q(x,t)=q(h(x,t)) [see (2.6)].

Assuming a sewer link with inflow q_u and outflow q, the time lag is translated into an according number of time steps for a discrete-time representation. In addition to the time lag, a linear first-order dynamic element may be included to increase the modelling accuracy for the link element. More precisely, the first-order system is introduced to take into account storage phenomena in the sewer and the dispersion of the flow, that is, the fact that the wave attenuates, lowering the peak flow depth and rate as it moves through the network and increasing the flow duration at any location. This leads to the following discrete-time dynamic equation:

$$q(k+1)=(1-T/\tau)q(k)+(T/\tau)q_u(k) \qquad (2.10)$$

where:
- $k = 1, 2, \ldots$ is the discrete time index.
- q is the outflow from the time-lag link.
- τ is the time constant of the linear first-order system that may be estimated experimentally.
- q_u is the sum of inflows to the time-lag link. Thus, if m is the total number of inflows into the sewer stretch, we have

$$q_u(k)=\sum_{j=1}^{m} q_{u,j}(k-\kappa_j) \qquad (2.11)$$

where κ_j is the time delay of inflow j.

Because the sewer stretch has a limited flow capacity q_{max}, we have the constraint

$$q_u(k) \le q_{max}. \qquad (2.12)$$

It should be noted that due to consideration of the backpressure phenomena, the inflows in hydrodynamic links do not need to be restricted explicitly by the flow capacity of the sewer.

2.2.2 Reservoirs

The reservoirs are modelled through the continuity equation

$$\frac{dV}{dt} = \dot{V} = u_{in}(t) - u(t) - q_{over}(t) \qquad (2.13)$$

where:
- $V(t)$ is the currently stored volume in the reservoir (in m^3).
- $u_{in}(t)$ is the inflow to the reservoir (in m^3/s).
- $u(t)$ is the outflow from the reservoir (in m^3/s).
- $q_{over}(t)$ is the overflow from the reservoir (in m^3/s).

The water level h that corresponds to the volume V is given by an inverse storage function H(V).

If an overflow weir is present and the water height exceeds the height of the weir, the reservoir will have an overflow (Figure 2.3). Under the assumption that there is no spillback downstream of the weir, the overflow q_{over} over a weir is given (Duncan et al., 1970; Papanikas, 1981) by

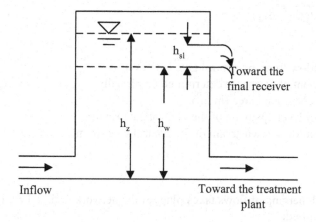

Figure 2.3. Reservoir with overflow capability.

$$q_{over} = \begin{cases} 0, & \text{if} \quad h_z - h_w \leq 0 \\ \frac{2}{3}\mu_p\sqrt{2g}(h_z - h_w)^{3/2}l_w \text{ (Poleni formula)}, & \\ & \text{if } h_z - h_w \leq h_{sl} \\ & \text{else} \\ \mu_T\sqrt{2g}(h_z - h_w - h_{sl}/2)^{1/2}A \text{ (Toricelli formula), else} \end{cases} \qquad (2.14)$$

where:
- h_w is the height of the weir (in m).
- h_{sl} is the height of the slot over the top of the weir (in m).
- l_w is the length of the weir (in m).
- μ_p is an overflow coefficient (Poleni formula).
- μ_T is a coefficient for the flow under pressure from the slot (Toricelli formula) ($\mu_T = (2/3)\mu_p\sqrt{2}$).
- A is the area of the slot (in m²) (A = $h_{sl}\, l_w$).

2.2.3 Control Gates

For the flow control in a combined sewer network, control gates are used. The control gates are placed at the end of sewer stretches or at the low points of the reservoirs.

The outflow from a control gate depends either on the water level upstream and downstream of the control gate or only on the water level upstream of the control gate when there is no backpressure. The outflow from a control gate is given by the nonlinear relationship

$$u_{un} = \mu f \sqrt{2g(h_u - h_d)} \qquad\qquad (2.15)$$

where:
- u_{un} is the outflow from the control gate (in m^3/s).
- μ is a coefficient of discharge determined empirically.
- f is the orifice's ground area (in m^2).
- h_u is the water level upstream of the control gate (in m).
- h_d is the water level downstream of the control gate (in m).

2.2.4 Nodes

The propagation and merging of flows takes place at the network nodes. Two types of nodes can be identified:

- Propagation: Nodes with one incoming link and one outgoing flow for connecting sewers with different geometry.
- Merging: Nodes where more than one incoming flows merge to one outgoing flow.

According to the continuity conditions, the following equations hold:

- For propagation nodes

$$u_{in} = u_{out} \qquad\qquad (2.16)$$

where u_{in} is the inflow to the node and u_{out} is the outflow from the node.
- For merging nodes

$$u_{in,1} + u_{in,2} + ... = u_{out} \qquad\qquad (2.17)$$

where $u_{in,i, ...}$, $i = 1, 2, ...$, are the incoming flows and u_{out} is the outgoing flow from the node.

2.2.5 External Inflows

The external inflows d_i occur at specific locations of the network. An external inflow may be present either directly at reservoirs or at the stretches of the sewer network.

2.2.6 Treatment Plants

Treatment plants finally receive all the water that entered the sewer network and did not exit from the overflows. The inflow r_i to a treatment plant i should not exceed the plant's flow capacity $r_{i,max}$:

$$r_i \le r_{i,max}. \qquad\qquad (2.18)$$

2.3 Simulation Tools for Sewer Networks

This section presents a survey of some of the existing commercially or semi-commercially available software packages used for the simulation of wastewater systems. Most of these software products model in an integrated way the sewer network, the wastewater treatment plant, and the receiving waters, and they also take into account the pollutants in the system. A detailed description of the simulation software package *KANSIM* used in this work is given in Section 2.4. It should be noted that in the version used in this study, KANSIM models only the water flow dynamics in the wastewater system.

The software packages reviewed in this section are MOUSE, SIMBA, HYDROWORKS, FLUPOL, MOSQITO, KOSIM, SYNOPSIS, SWMM, WEST, ICS, and SUPERLINK.

The package *MOUSE* (Modelling of Urban Sewers) is one of the most widely used tools for modelling of urban sewer systems. In this package there are different modules for describing the hydrological effects and the hydraulic performance of sewer systems (Hernebring *et al.*, 2002). In the general hydrological model *MouseRDII* (Rainfall Dependent Inflow and Infiltration) the hydrological processes are described, taking into account both the fast response component from impervious areas and the slow response component caused by infiltration into the sewer system from the surrounding soil, and in the hydrodynamic model *MousePIPE* the hydraulics of the sewer system are described using the Saint-Venant equations. The sewer flow quality model *MOUSETRAP* based around the sewer model MOUSE was developed by a consortium including representatives of the Danish Hydraulic Institute, Water Quality Institute, Water Research Center (WRC), and others (Schütze *et al.*, 2002). The MOUSETRAP package has a surface runoff quality module, a sediment transport module, an advection-dispersion module and a water quality module in which the biochemical processes are modelled (Ashley *et al.*, 1999; Schütze *et al.*, 2002). In Hernebring *et al.* (2002) the MOUSE package was used to model the Helsingborg sewer network in Sweden; while a MOUSE ONLINE system was installed and connected to a SCADA (Supervisory Control and Data Acquisition) system (Katebi *et al.*, 1999) of the Öresundsverket wastewater treatment plant of the Helsingborg municipality and performs a series of tasks automatically within certain intervals, such as collecting data from the SCADA system, generating forecasts of boundary conditions, transferring measured and forecast data to the MOUSE system, performing one or more MOUSE simulations, and possibly performing the extraction of data/information from MOUSE ONLINE back to the SCADA system. In Jacopin *et al.* (2001) the MOUSE model was used for modelling the Bourgailh and Périnot basins of the Urban Community of Bordeaux (France) in the frame of a project with the main scope of developing new control schemes for detention basins during rainfall events in order to limit the flooding risk, especially for heavy rainfall events, and, at the same time, to promote the solids settling process in basins, especially for light and medium events.

SIMBA is a simulation platform running on top of MATLAB/SIMULINK; it was developed at Ifak (Institut für Automation und Kommunikation) e.V.,

Magdenburg (Germany) (Rauch *et al.*, 2002; Schütze *et al.*, 2002). In SIMBA there are modules for simulating water flow and quality processes in the sewer systems (*SIMBA Sewer*), treatment plant, and rivers. In SIMBA the use of the general-purpose simulation environment MATLAB/SIMULINK allows the user to add her own modules, whereas the control library with defined control blocks, which is available in SIMBA, makes this package a convenient tool for control and optimisation of the overall system performance. In Frehmann *et al.* (2002) and Erbe *et al.* (2002) in the urban catchment area called "Schattbach" in the southeast of Bochum (Germany), the conveyance of the discharge and the pollutants within the sewer network and the facilities for rainwater treatment was conducted by use of the program SIMBA Sewer, whereas the SIMBA package was used to calculate the biological wastewater treatment plant on the basis of ASM1 (Activated Sludge Model No. 1). In Erbe *et al.* (2002), the SIMBA Sewer and SIMBA were used for the integrated modelling of the wastewater system of Odenthal (Germany), whereas in Masse *et al.* (2001) the SIMBAD model was used for dynamic wastewater treatment plant modelling of the sewerage system of Grand-Couronne, Normandy (France).

The HYDROWORKS package has been developed by Wallingford Software (UK) and Anjou Reserche (France). HYDROWORKS is an integrated package, that has modules for sewer systems and rivers (Schütze *et al.*, 2002). The hydraulic model of HYDROWORKS is based on the full Saint-Venant equations. HYDROWORKS also includes a real-time control (RTC) module and a quality module. A mass-conservation approach is used for modelling the transport of suspended sediment and dissolved pollutants, whereas no physical or biochemical processes are modelled in HYDROWORKS (Ashley *et al.*, 1999; Schütze *et al.*, 2002). In Zug *et al.* (2001) this simulation package (hydraulic, quality, and RTC module) has been used for modelling a section of the Grand-Nancy (France) sewerage system, the Gentille tank, whereas in Masse *et al.* (2001) the sewer network of Grand-Couronne, Normandy (France) is modelled by using this simulation model.

The *FLUPOL* package was developed by Agence de l'Eau Seine-Normandie (AESN), the Syndicat des Eaux d'Ile-de-France (SEDIF), and the Compagnie Générale des Eaux (CGE) (Schütze *et al.*, 2002). The hydraulic model of FLUPOL uses three flow-routing models (Muskingum-Cunge, kinematic, and diffusive wave), whereas the quality model simulates pollutant transport (without dispersion), sedimentation, and erosion of sediments in sewers, but no biochemical processes are modelled (Ashley *et al.*, 1999).

The *MOSQITO* sewer flow quality model was developed by Wallingford Software, based on research carried out by HR Wallingford and WRC. MOSQITO simulates surface runoff, pollutant transport (without dispersion), sedimentation, washoff, and sediment transport, but no biochemical processes (Schütze *et al.*, 2002). In the quality module of HYDROWORKS, modelling developments of MOSQITO and FLUPOL are included.

The package *KOSIM* is a sewer flow-quality model that has been developed and extended by several researchers and is used in engineering practice as well as in numerous research projects (Schütze *et al.*, 2002). The flow on subcatchments is modelled by Nash cascades, whereas the flow between subcatchments is modelled by translation and addition of inflow from the subcatchments. In KOSIM up to six pollutants can be modelled once the pollutant-specific parameters are defined and they are routed through the system, where they are assumed to mix completely and without any interactions.

The software tool *SYNOPSIS* (software package for synchronous optimization and simulation of the urban wastewater system) is an integrated simulation (capable of simulating the water and quality processes in sewer system, treatment plant, and receiving water) and optimization tool. Three existing software packages are selected for implementation in SYNOPSIS (Schütze et al., 2002). The sewer system module of SYNOPSIS consists of the program package EWSIM, which is an extended (research) version of the commercially available package KOSIM. For the treatment plant the Lessard and Beck's treatment plant model was selected and some modifications and extensions were made in this model within SYNOPSIS. As a river simulator within SYNOPSIS, the DUFLOW shell program has been selected. Besides these modules, a number of auxiliary routines have been implemented in SYNOPSIS, such as modules computing external inputs to the simulation model, routines performing the transformation of variables at the interfaces between the simulation submodules, and modules providing a variety of values that are available as options for the definition of the objective function for the optimization. SYNOPSIS has the ability to perform control actions with its optimization module. Three optimization routines, –the controlled random search, a genetic algorithm, and Powell's method for local optimization– have been implemented in the optimization module.

The program package SWMM (Storm Water Management Model) developed by the U.S. Environmental Protection Agency, models flow and pollutants in sewer systems (Ashley et al., 1999; Schütze et al., 2002). Its transport module does not allow for simulation of backwater effects and pressure flows to be considered. For these flow processes, EXTRAN, an extended transport model, is available.

The software WEST (World Engine for Simulation and Training) developed by Hemmis n.v., University of Gent and Epas n.v. in Belgium (Rauch et al., 2002; Schütze et al., 2002) is a general simulation environment for computing the flow dynamics in a network of interlinked elements. With its different modules for sewers, wastewater treatment plants, and rivers, and its controller that allows several options of control, WEST can be used for the integrated simulation and control of the complete urban wastewater system. In Meirlaen et al. (2002), an integrated modelling of the urban wastewater system of the town of Tielt has been accomplished by implementing surrogate models of the three subsystems (sewer system, wastewater treatment plant, receiving water) within the single software platform WEST.

The *ICS* (Integrated Catchment Simulator) program developed by Danish Hydraulic Institute (DHI) and the Water Research Center in UK (WRC) is a graphical interface for setting up and running integrated models with feedforward/feedback of information and includes existing modules for sewer (MOUSE), wastewater treatment plants (STOAT), and rivers (MIKE11) (Rauch et al., 2002).

The *SUPERLINK* model is a hydraulic numerical model developed for complex looped sewer or channel systems using an implicit scheme with emphasis on the stability and the computational speed (Ji *et al.*, 1996). The Saint-Venant flow equations are used to describe one-dimensional unsteady nonuniform flow under both free-surface and pressurized flow conditions (SUPERLINK is one of the fastest sewer models available to solve the full Saint-Venant equations). The SUPERLINK model setup has the convenience of including a pollutant transport and water quality model. Besides the simulator submodel, SUPERLINK has a controller submodel, which is separated from and interacts with the simulator submodel via alternating calls. In Ji *et al.* (1996), the SUPERLINK model was used for a part of a large sewer network in the city of Winnipeg (Canada), and the simulation results showed that the model achieves a significant improvement in computation speed compared with the USEPA EXTRAN model and can still maintain similar accuracy. [The *USEPA EXTRAN* model solves the Saint-Venant flow equations using a completely explicit finite difference scheme and offers a node-link setup scheme aiming to solve for flow in links and head at nodes (Ji *et al.*, 1996; Duchesne *et al.*, 2001).]

2.4 Simulation Program – KANSIM

In this section, the simulation program KANSIM is presented based on the accurate model of the sewer network that was previously described. This program is written in C and is using three input files – one for describing the sewer network (the topology and the geometric characteristics of the sewer network's elements), one for the definition of the parameters that are needed for the simulation, and one that contains the external inflows. A detailed description of this program is given in Messmer (1998). KANSIM was initially considering water flow only, but has been recently extended to also consider the pollutant dynamics.

2.4.1 Link Elements

a) Hydrodynamic Link Elements

In the simulation program KANSIM, spatial discretization is introduced for the numerical solution of Equation (2.1). A geometrically uniform piece of sewer (constant slope and profile) is modelled as one link divided in n segments of length Δx. Figure 2.4 shows this segmentation. For realistic modelling of the flow, the segments have to be chosen sufficiently short. The spatial derivative $\partial q / \partial x$ is replaced by the approximation $\Delta q / \Delta x = (q_i - q_{i-1}) / \Delta x$. After substitution in Equation (2.1) and some manipulations, the following ordinary first-order differential equation is obtained per segment i (continuity equation):

$$\frac{dV_i}{dt} = \dot{V}_i = q_{i-1} - q_i \tag{2.19}$$

where $V_i = F(h_i)\Delta x$ represents the volume stored in segment i, h_i being the water level in segment i (Figure 2.4). The differential Equation (2.19) is solved using numerical methods (see Section 2.4.2).

By replacing $\partial h/\partial x$ with $(h_{i+1} - h_i)/\Delta x$ and I_S with $\Delta h_S/\Delta x$ (see Figure 2.4) and by substituting these relations in Equation (2.5) the following equation is obtained:

$$q_i = \frac{kR(h_i)^{2/3}F(h_i)}{\sqrt{\Delta x}}\sqrt{\Delta h_S + h_i - h_{i+1}}. \qquad (2.20)$$

The variable h_i is a direct function of the volume:

$$h_i = H(V_i) \qquad (2.21)$$

where H(V) is the inverse function of $F(h)\Delta x$. The constant Δh_S is the difference of the sewer bottom elevation over the length of one segment. As water level values are available only for the beginning and the end of a segment, a parameter α is introduced to take into account the upstream and the downstream water levels, and a variable w_i is used that can be regarded as a resistance value for the section between the segments i and i+1 and is calculated from Messmer (1998)

$$w_i = \sqrt{W(h_i)^2 \alpha + W(h_{i+1})^2 (1-\alpha)}\sqrt{\Delta x} \qquad (2.22)$$

where

$$W(h_i) = \frac{1}{kR(h_i)^{2/3}F(h_i)} \qquad (2.23)$$

where α $(0 < \alpha < 1)$ is the partition factor. Thus, the flow between segments can be calculated by the algebraic equation

$$q_i = \sqrt{\Delta h_S + h_i - h_{i+1}}/w_i. \qquad (2.24)$$

b) Time-Lag Links

For the time-lag links in the simulation program KANSIM, the time lag x/c (see Section 2.2.1) has a constant value. This value is translated into a corresponding number of time steps according to the relation

$$\kappa = entier(l/(c_N T) + 0.5) \qquad (2.25)$$

where:

- l is the length of the sewer (in m).
- T is the discrete time interval of the simulation (in s).
- c_N is the nominal velocity ($c_N = \partial q/\partial h\big|_{h_N} \big/ \partial F(h)/\partial h\big|_{h_N}$) that corresponds to the nominal water level h_N, the water level that corresponds to the half of the sewer's height.

The function entier(x) expresses the truncation of x to the next lower integer. For the calculation of flow, Equations (2.10) and (2.11) are used.

2.4.2 Reservoirs

Equation (2.13), which is used for mathematical modelling of reservoirs, may be solved numerically using the Euler-backward method, which is better with respect to stability, or the trapezoidal rule (or 1-step Adams-Moulton method), which is better with respect to accuracy (Lambert, 1991). In fact these methods are used in the program KANSIM for solving all the differential equations of the form $\dot{V} = \Delta q$, that is, also for hydrodynamic links. The Euler-backward method approximates the differential equation by the following difference equation:

$$V(k+1)=V(k)+T\Delta q(k+1) \tag{2.26}$$

whereas the trapezoidal rule employs the following difference equation:

$$V(k+1)=V(k)+0.5T\,(\Delta q(k)+\Delta q(k+1)) \tag{2.27}$$

where k is the discrete time index and $\Delta q(k)$ and $\Delta q(k+1)$ are calculated, for example, for hydrodynamic links from Equation (2.24). Both methods are implicit, as V(k+1) is needed on both sides of the difference equation. Therefore, at every time step, (2.26) [or (2.27)] is solved iteratively for V(k+1) by use of Newton iterations.

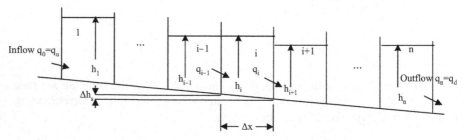

Figure 2.4. Segmentation of a hydrodynamic link.

2.4.3 Control Gates

In the simulation program KANSIM, the flow through the control gates is not calculated by Equation (2.15); instead, the control gates are considered as link elements that produce additional flow resistance w_s when control gates are placed between a reservoir and a hydrodynamic link or between hydrodynamic links. The additional flow resistance is taken into account in the calculation of flow over a node (see Section 2.4.4 below); that is, the variable w_d in Equation (2.30) is accordingly increased and is equal to $(w_d^2 + w_s^2)^{1/2}$. The variable flow resistance w_s, which depends on the opening height of the gate and the water levels upstream and downstream of the gate, is calculated from the relation (Messmer, 1998)

$$w_s = \begin{cases} \dfrac{1}{(2.2723 + 0.0802n_{hu})ab} & \text{if } n_{hu} < 3.6 \\ \dfrac{1}{(2.4672 + 0.0259n_{hu})ab} & \text{else} \end{cases}$$
(2.28)

where:
- a is the opening height (in m) [the opening height a has a minimum value a_{min} so as to avoid a division by zero in equation (2.28)].
- b is the width of the gate (in m).
- $n_{hu} = h_u/a$.

The control gates are actuators of a local flow control loop. The local control loop regulates the flow from the control gate to be near a set point as long as the physical conditions permit. The set point of this flow can have a constant value to be read from an input file, or a variable value at every time step of the simulation if some control strategy is programmed by the user. The flow cannot take the set point value if:
- the upstream water level (pressure) is too low or backpressure prevents the flow from reaching the desired flow even with fully opened gate, or
- there is a minimum admissible gate opening, leading to a flow that exceeds the desired value beyond a certain upstream water level.

2.4.4 Nodes

In the simulation program KANSIM, the types of nodes presented in Section 2.2.4 are considered in addition to overflow nodes. Overflow nodes have one incoming flow and two outgoing flows. One outgoing flow is moving toward the next element of the sewer network (along the direction to the treatment plant), and the other outgoing flow over a weir is directed toward the final receiver. The overflow nodes have at least one adjacent link with a storage characteristic, reservoir, or hydrodynamic link element. For overflow nodes the following equation holds:

$$u_{in} = u_{out,1} + q_{over}$$
(2.29)

where u_{in} is the inflow to the node, $u_{out,l}$ is the outflow that is moving toward the following element of the sewer network, and q_{over} is the overflow given by Equation (2.14).

The merging nodes are considered having two incoming and one outgoing flow. To calculate the flow at the nodes and in analogy to the flow calculation between segments of hydrodynamic links [Equations (2.22), (2.23), (2.24)], upstream w_u and downstream w_d flow resistances of the node are considered (Messmer, 1998).

When the adjacent links of a node are hydrodynamic link elements, w_u is the flow resistance of the last segment ($\alpha \cdot \Delta x_\mu$) of the incoming link element μ, and w_d is the flow resistance of the first segment (($1-\alpha$)$\cdot\Delta x_v$) of the outgoing link element (where α is the partition factor defined in Section 2.4.1 and Δx_i is the length of segment i). In case of a time-lag link as an outgoing link, w_d is equal to zero (Messmer, 1998).

By introducing an auxiliary intermediate water level h_z (Figure 2.5) at the node, the calculation of the flow can be expressed in the following general form (Messmer, 1998):

$$h_z = h_d + (1-\alpha) \cdot \Delta h_{s,v} + (u_{out,\lambda} w_d)^2 \tag{2.30}$$

and

$$u_{in,\lambda} = \sqrt{h_u + \alpha \cdot \Delta h_{s,\mu} + h_z}/w_u \tag{2.31}$$

where $u_{out,\lambda}$ is the inflow of the outgoing element v of node λ and $u_{in,\lambda}$ is the outflow of the incoming element μ of node λ. Equation (2.31) is applied to both incoming link elements in the case of merging nodes. It is not applied, however, in cases of time-lag links or control gates with operative flow control loop as incoming elements, as in these cases $u_{in,\lambda}$ is directly given. For the overflow elements, Equation (2.30) holds true only for the straightly leaving flow $u_{out,i,l}$, whereas the flow over the weir is given by Equation (2.14).

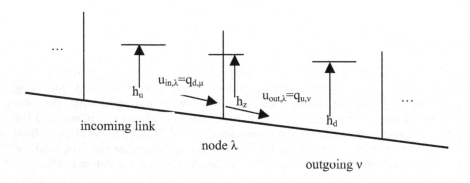

Figure 2.5. Variables at a node.

2.4.5 External Inflows

In the program KANSIM, external inflows are input data to be read from the input file.

2.4.6 Treatment Plants

In the program KANSIM, the flow capacity $r_{i,max}$ of the treatment plant is given in the input file that describes the characteristics of the sewer network.

2.5 Simplified Model of Sewer Networks

In the simplified model of the sewer network some simplifications are introduced with respect to the accurate model that was described in Section 2.2 so as to keep the computational effort for control within reasonable levels. More precisely, in the simplified model:

- For modelling the flow in the sewers, only time-lag link elements are used, and hence the simplified model of the sewer network cannot describe backpressure effects. The connection of sewer links with different geometry is effectuated without the use of propagation nodes, whereas only merging nodes are used in the simplified model.
- Overflows over weirs at reservoirs are calculated according to the reservoir current storage and storage capacity.

In the following subsections, the elements of the sewer network are described according to the simplified model.

2.5.1 Link Elements

Link elements (Figure 2.6a) model the flow process in the sewer stretch. These elements are necessary if the flowing time of waves along a sewer stretch significantly exceeds the sample time interval T. A real sewer stretch may be modelled by one single link or by a tandem connection of several link elements. The outflow from the link element is given by Equations (2.10) and (2.11) and the total inflows to the link are restricted according to Equation (2.12).

2.5.2 Reservoirs

Reservoirs are considered as storage elements with an inflow u_{in}, an outflow u, and an overflow q_{over}, (Figure 2.6b). The discrete-time equation describing the filling and emptying of a reservoir i, is

$$V(k+1) = V(k) + T(u_{in}(k) - u(k) - q_{over}(k)) \qquad (2.32)$$

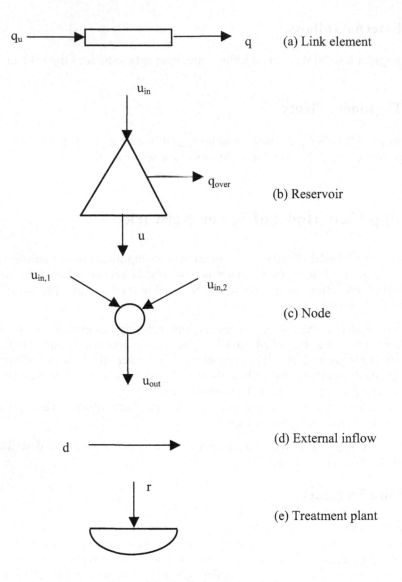

Figure 2.6. Symbols of network elements.

where:

- T is the discrete time interval.
- $k = 0,1,...$ is the discrete time index.
- $V(k)$ is the reservoir storage at time kT.
- $u_{in}(k)$ is the sum of inflows (from elements of the sewer network that are located upstream of the reservoir but also external inflows) over the period $[kT, (k+1)T]$.

- $q_{over}(k)$ is the overflow of the reservoir i over the period $[kT, (k+1)T]$. Instead of Equation (2.14) of the accurate model, it is assumed that an overflow occurs if $V(k) \geq V_{max}$, where V_{max} is the storage capacity of the reservoir. A parameter d_0, to be estimated empirically, is introduced in the simple relationship in order to best fit the linear equation to the overflow values obtained by the nonlinear relation (2.14). Thus, overflow is given by

$$q_{over}(k) = \begin{cases} 0 & \text{if } V(k) \leq V_{max} \\ d_0[V(k) - V_{max}]/T & \text{else.} \end{cases} \tag{2.33}$$

- $u(k)$ is the controllable outflow from the reservoir over the period $[kT, (k+1)T]$. The controllable outflows $u(k)$ of all reservoirs are the input variables to be calculated by the central controller. As in the simplified model of the sewer network some processes cannot be modelled exactly; the outflows $u_i(k)$ should be selected from an admissible control region so as to take indirectly into account certain physical limitations:

$$u_{min} \leq u(k) \leq u_{max}(V(k),k) \tag{2.34}$$

where $u_{min} \geq 0$ and

$$u_{max}(V(k),k) = \min\{u_{cap}, u_{un}(V(k)), u_o(V(k),k)\} \tag{2.35}$$

where:
- u_{cap} is the outflow capacity of the downstream sewer stretch.
- $u_{un}(V(k))$ is the outflow from the reservoir that occurs when the outlet gate is completely opened. For example, when the water level downstream of the reservoir is near the height of the orifice and the water level in the reservoir is greater than twice the height of the orifice, the following relationship can be used (Bretschneider *et al.*, 1982) [This relationship is a simplified form of the Equation (2.15) that is used in the accurate model of the sewer network]

$$u_{un} = c_0 f \sqrt{h_u} \tag{2.36}$$

  where:
  - c_0 is an empirical parameter (in $m^{0.5}/s$),
  - f has the same meaning as in Equation (2.15), and
  - h_u is the water level in the reservoir (see 2.21).
    Equation (2.36) can be directly replaced in (2.32) in cases of reservoirs with uncontrollable outflow.
- $u_o(V(k),k)$ is equal to $V(k)/T$ plus the corresponding external inflow to the reservoir (if there is any). This upper value guarantees that $V(k) \geq 0$.

2.5.3 Nodes

The nodes (Figure 2.6c) with outflows u_{out} of the simplified model of the sewer network correspond to the merging nodes of the accurate model of the sewer network (2.17).

2.5.4 External Inflows

External inflows (Figure 2.6d) of the simplified model of the sewer network are described in the same way as in the accurate model of the sewer network.

2.5.5 Treatment Plants

The treatment plants (Figure 2.6e) of the simplified model of the sewer network are described in the same way as in the accurate model of the sewer network. The inflow r to the treatment plant is constrained by Equation (2.18).

2.5.6 Integrated Simplified Model of the Sewer Network

A particular network can be built upon the introduced elements. The whole flow process may be considered to have a vector input $\mathbf{u}$ including all controllable reservoir outflows, a disturbance vector $\mathbf{d}$ including all external inflows, and a state vector $\mathbf{x}$ including all reservoir storages and link outflows. Then, all model Equations (2.10), (2.11), (2.17), (2.32), and (2.33) may be expressed in the following general form:

$$\mathbf{x}(k+1) = \mathbf{f}[\mathbf{x}(k), \mathbf{x}(k-1),...,\mathbf{x}(k-\kappa_x), \mathbf{u}(k), \mathbf{u}(k-1),...,$$

$$\mathbf{u}(k-\kappa_u), \mathbf{d}(k), \mathbf{d}(k-1),...,\mathbf{d}(k-\kappa_d)]. \tag{2.37}$$

Similarly, we have the control constraints (2.18), and (2.34) in the form

$$\mathbf{u}_{min} \leq \mathbf{u}(k) \leq \mathbf{u}_{max}(\mathbf{x}(k),k). \tag{2.38a}$$

and the constraints (2.12):

$$\sum_{\kappa=1}^{\kappa_u} \mathbf{A} \cdot \mathbf{u}(k-\kappa) \leq \mathbf{c}(\mathbf{x}(k),k) \tag{2.38b}$$

where the matrix $\mathbf{A}$ consists of zeros and ones. Equation (2.38b) is derived by substituting Equation (2.10) in (2.11) and keeping the controllable outflows of the reservoirs on the left-hand side of the equation that results after the substitution. The other inflows into the sewer (external inflows, link outflows, and uncontrollable reservoir outflows) are subtracted from $\mathbf{q}_{max}$ on the right-hand side of the equation. As the outflows from the sewers are state variables [Equation (2.10)] and the uncontrollable reservoir outflows depend on the state variables [Equation (2.36)], the resulting right-hand side of Equation (2.38b) is state-dependent.

Chapter 3
Flow Control in Sewer Networks

3.1 Control Objectives

The development of a control system for combined sewer networks has as a goal the protection of the quality of waters that receive the outflows of the networks, as was mentioned in Chapter 1. To this end, the main task of the control system is the *minimization of overflows* for any rainfall event. This can be achieved by:

- Using all available storage space before allowing an overflow to occur somewhere in the network. Moreover, if, due to strong rainfall, overflows are unavoidable, they should be distributed as homogeneously as possible over time and over the network reservoirs in order to minimize their polluting impact. However, if there are storage elements without overflow capability (no overflow weirs), the avoidance of overload of these storage elements is of primary importance.
- Emptying the network as soon as possible (via full exploitation of the treatment plant's inflow capacity) so as to provide free storage space for a possible future rainfall.

Along with these main objectives, some secondary operational objectives are expected to be sufficiently addressed by the control actions:

- The distribution of the current total storage volume among the reservoirs should be according to predefined portions (that are typically proportional to the reservoirs' storage capacities); in other words, at any time but mainly during the emptying phase, free storage space should be available in each reservoir to be used in case of a possible future rainfall.
- Limitation of abrupt changes of outflows, in order to account for throttle gate inertia.

3.2 Multilayer Control System

The flow control of a sewer network, which usually covers large areas and consists of a high number of reservoirs and sewers, calls for the consideration of the network as a whole, as the isolated control of each reservoir cannot lead to full utilization of the total available storage volume. A control structure that consists of three control layers (Papageorgiou and Mevius, 1985) (adaptation, optimization, and direct control, see Figure 3.1) may be used for the control of the sewer network. The formulation of the central control problem for the optimization layer on the basis of an accurate mathematical model of the sewer network, such as the one described in Section 2.2, may lead to long computation time for the solution of the associated dynamic optimization problem in real time for large-scale networks. Therefore, a simplified model of the sewer network, such as the one of Section 2.5, is used as a basis for mathematical modelling of the control problem. The solution of the control problem in the optimization layer delivers the trajectories for the outflows (control variables) and the storages (state variables) of the reservoirs that are used as reference trajectories for the decentralized direct control of the subnetworks (reservoirs).

In the following subsections, some more details related to the specific control layers of the multilayer control system are presented (Figure 3.1):

Adaptation Layer

In this layer at time step k_0 the prediction of the external inflows is performed for k = $k_0,..., k_0+K-1$, where K is the optimization horizon. Measurements of the rainfall may be used in the prediction model that calculates the predicted inflows to the network. The more accurate the prediction, the better the control results, though the sensitivity of the control to inaccurate inflow predictions is limited as repeated solution of the control problem in real time is performed with updated inflow predictions (see Section 4.4). In this layer also the current values of the state variables are estimated.

Optimization Layer

This layer calculates the control and state trajectories for k = $k_0,..., k_0+K-1$, for the overall simplified network. The control problem is solved using the initial values of the state variables and the trajectories of the expected external inflows that are provided by the adaptation layer. For the solution of the control problem, different methods can be used, like expert systems, fuzzy control, and other methods (see Chapter 1). The solution of the control problem corresponds to the outflows and storage trajectories of the reservoirs, which are used as reference trajectories for the direct control layer.

Distributed Direct Control Layer

The optimal outflows and storages of the reservoirs that are calculated in the optimization layer are used as reference trajectories for the decentralized control of each subnetwork (reservoir). In each of these subnetworks there are local control systems that operate in a closed-loop structure and move the control devices, for example the control gates, so as to keep the real quantities near their reference trajectories. The local control systems are autonomous in the sense that the control devices operate using measurements from the subnetwork only. The coordination between subnetworks is done through the reference values that are provided from the optimization layer, taking into account the overall network.

If some significant deviation of the predicted inflow from the real inflow occurs or if the local controllers cannot follow the reference trajectories, the whole procedure is repeated with updated inflow predictions and new estimates of the initial state.

This control structure is:

- Flexible, because of its modular character, with well-defined interfaces allowing application of different methods at different layers.
- Reliable with respect to failures of system components due to its decentralized structure. Thus, as the local control strategies are implemented in microprocessors, if the central computer has a failure or communication channels to the local systems are cut, the local systems may consider as reference trajectories some suitable and safe (though non-optimal) stored values. Also, if a local microprocessor fails, the central computer defines new reference values for the microprocessors that are still functioning, taking into account the failure.
- Efficient due to the real-time operation that is automatically adaptable to changing exogenous conditions, and also due to the global consideration of the overall network.

This book is focused on the analysis and study of the control methodologies used in the optimization layer.

3.3 Studies of Water Resource Systems

The development of optimization techniques for planning, design and management of complex water resources systems has been the subject of many investigations throughout the world. The choice of the method to be used for the optimization depends on the characteristics of the reservoir system being considered, on the availability of data, and on the specific control objectives and constraints. Many researchers in the field have considered methods such as linear programming, dynamic programming, nonlinear programming, linear-quadratic control, genetic algorithms, and combinations of these methods.

Linear programming is a very powerful and easy-to-use form of optimization. For example, in Gutman (1986) a linear programming regulator operates in an open-loop optimal feedback (or model-predictive) structure to control in real time a double water tank laboratory process and the water level of a hydroelectric power station reservoir leading to satisfactory performance. For sewer network control, linear programming is used in Bradford (1977) for the development of a control

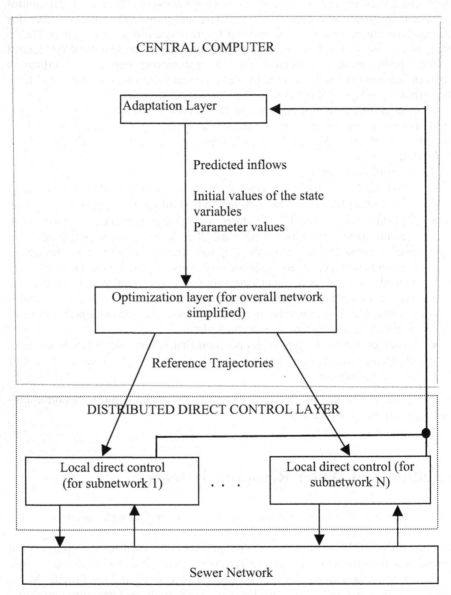

Figure 3.1. Hierarchical control structure for sewer networks.

algorithm for automatic control of detention storage in a large-scale combined sewer system and in Nelen (1994) for the real-time control of urban drainage systems, whereby the nonlinear programming problem is replaced by a succession of linear programming problems. Linear programming is most efficient for problems that can be expressed in linear terms.

Dynamic programming has been used extensively in the optimization of water resources systems, as the nonlinear and stochastic features, which characterize a large number of water resources systems, can be readily translated into a dynamic programming formulation. However, when dynamic programming is applied to multiple reservoir systems, the usefulness of the technique is limited, as the computer memory and computation time requirements are exponentially increasing. In such cases, dynamic programming can be applied only if the complex problems with the large number of variables are decomposed into a series of subproblems, which are solved recursively. For example, in Zessler and Shamir (1989), progressive optimality, an iterative dynamic programming method, is used for determining optimal operation of a water supply system. The algorithm is solved iteratively over the time steps and network subsystems, and converges to the optimum from any initial solution. Discrete differential dynamic programming (an iterative technique in which the recursive equation of dynamic programming is used to search for an improved trajectory among the discrete states in the neighborhood of a trial trajectory) is used by Heidari *et al.* (1971) for a four-unit, two-purpose water resources system and in Meredith (1975) to determine the optimal operation policy of a multiple-purpose multiple-reservoir system, whereas in Murray and Yakowitz (1979) constrained differential dynamic programming is applied to multireservoir control problems. Combinations of linear programming and dynamic programming were also proposed for water system control; for example, in Grygier and Stedinger (1985) such a combination is used to optimize the operation of multireservoir hydrosystems as well as in Yeh and Becker (1982) for the development of practical procedures for the analysis of multiple-purpose, multiple-facility reservoir systems to guide real-time decisions concerning the optimal operation of the system. In the context of sewer network control, dynamic programming has been used for optimizing the design of drainage systems (Robinson and Labadie, 1981), for designing the least expensive network of sewers that will drain water from a number of discrete sources (Walters, 1985), for designing the least-cost drainage networks, which include storage elements (Froise and Burges, 1978), and for control of the combined sewer network of the city and county of San Francisco (Labadie *et al.*, 1980).

Nonlinear programming offers a more general mathematical formulation than linear and dynamic programming and can effectively handle nonlinear objective functions and nonlinear constraints. Methodologies based on Pontryagin's maximum principle are used for solving the scheduling problem of hydroelectric power plant chain (Sakr and Dorrah, 1985), for solving the problem of the most economical operation of hydraulic plants in electric power systems (Hano *et al.*, 1966), for determining the operation of a hydro-steam generating system for the minimum generating costs (Dahlin and Shen, 1966), and for the optimal control of water supply networks (Nielsen and Ravn, 1985). Conjugate gradient algorithms are used for the optimal management of hydrostorage reservoirs (Sirisena and Halliburton, 1981), and for the optimal control of the complex multireservoir Mahaweli system in Sri Lanka (Mizyed *et al.*, 1992). In Chu and Yeh (1978), for the optimization of real-time operations of a single reservoir system, nonlinear duality theorems and Lagrangian procedures are applied, whereby the minimization of Lagrangian is carried out by a modified gradient projection technique along with an optimal stepwise determination technique, whereas in Saha and Khaparde (1978) the optimal scheduling of hydrothermal power systems is performed by a feasible direction algorithm. In Foufoula-Georgiou and Kitanidis (1988) an algorithm that combines elements of discrete dynamic programming (*i.e.*, discrete state space, backward stagewise optimization) with elements of constrained optimization (*i.e.*, nonlinear programming with equality constraints) is used for the optimal control of a multireservoir system. In Bell *et al.* (1973), optimal control theory is used for real-time automated control of combined sewers, whereas in Pleau *et al.* (1996), Méthot and Pleau (1997), Pleau *et al.* (2001) nonlinear programming is applied for the flow control of the Québec Urban Community sewer network; in Gelormino and Ricker (1994) a model-predictive control strategy that uses a mixed linear/quadratic objective function is applied to the Seattle metropolitan area in order to minimize combined sewer overflows. A solution algorithm developed for the sewer network control problem applying the discrete maximum principle has also been used (Papageorgiou, 1983, 1985; Papageorgiou and Mayr, 1985, 1988; Marinaki, 1995, 2002; Marinaki and Papageorgiou, 1997a, b, 1998, 1999).

Linear-quadratic control theory has been extensively applied in many fields, and a number of investigators have incorporated various aspects of linear-quadratic regulator theory in their proposed solutions of reservoir operations problems. For example, in McLaughin and Velasco (1990) a linear-quadratic control algorithm is applied to a system of hydropower reservoirs, in Garcia *et al.* (1992) a real-time compensation scheme for multipool canals is developed using linear-quadratic methods, and in Winn and Moore (1973) a multivariable feedback controller is used for the control of combined storm-sewer systems. A linear multivariable feedback regulator, designed using the linear-quadratic methodology, is used for the sewer network control problem (Messmer and Papageorgiou, 1992; Marinaki, 1995, 2002; Marinaki and Papageorgiou, 1996b, 1997a, b; Marinaki *et al.*, 1999, Marinaki, 2002).

Genetic algorithms have been proposed as a means of global optimization for a variety of engineering design problems. They mimic the natural genetic processes of evolution, deliberately keeping a range of good solutions to avoid being drawn into low-quality local optima. Genetic algorithms are robust methods for searching the optimum solution to complex problems, although they may not necessarily lead to the best possible solution. In Liu and Wu (1993), a technique combining analytical and knowledge models is proposed for modelling and control of large-scale water distribution systems, in Lee and Ellis (1996) genetic algorithms are applied to a simple network location problem, and in Wardlaw and Sharif (1999) they are applied to a four-reservoir, deterministic finite-horizon problem.

Conventional rule-based control and *fuzzy logic* for real-time flow control of sewer systems have also been proposed. Conventional rule-based control systems are based on a large number of rules, whereas control systems based on fuzzy logic combine the simple rules of an expert system with a flexible specification of output parameters. In Klepiszewski and Schmitt (2002), a comparison of a conventional rule-based flow control system with a control system based on fuzzy logic is effectuated for a combined sewer system and in Fuchs *et al.* (1997) a study was carried out for a part of the sewer system of the city of Flensburg using fuzzy logic for the real-time control of the sewer system.

3.4 The Pursued Approach

As already mentioned, the main focus of this monograph is the development, testing and application of a control system for the central flow control in combined sewer networks. The methods employed in this study for the solution of the central control problem are nonlinear optimal control (Chapter 4) and multivariable feedback control (Chapter 5). The selection of these two methods was made on the basis of their distinguished qualities. More specifically, nonlinear optimal control was selected because, as already mentioned in Chapter 1, it is potentially the most efficient approach, as it can directly take into account inflow predictions, and process nonlinearities and constraints; on the other hand, nonlinear optimal control calls for the development and implementation of a rather sophisticated computer code. Multivariable feedback control was selected (Chapter 1) because it may approximate the efficiency of nonlinear optimal control if properly designed but needs a much simpler computer code. Moreover, it should be emphasized that both approaches lead to quite straightforward general development procedures for virtually any sewer network, in contrast to other application-specific heuristic or rule-based approaches.

Some modifications and extensions of the methodologies used in the past for both employed approaches are introduced in order to increase their efficiency. More precisely, in the nonlinear optimal control approach, modifications (Papageorgiou and Marinaki, 1995) were introduced in the feasible-direction algorithm that mainly concern the constraints of the control variables, so as to reduce the computational effort that is needed until convergence. In the multivariable feedback control, appropriate modifications were introduced (Marinaki and Papageorgiou, 1996a) so as

- to take into account the time delays of the control and state variables,
- to develop a fully controllable linear model, and
- to develop a feedback controller that has additionally anticipatory behaviour.

Finally, with respect to a previous study (Marinaki, 1995), in this book:

- A more realistic simulation model, the program KANSIM presented in Section 2.4, is used for simulation testing and comparison purposes.
- The rolling horizon method (Papageorgiou, 1988, 1997) is used for the real-time application of nonlinear optimal control with updated inflow predictions and updated initial conditions.
- An investigation is performed concerning the sensitivity of the control efficiency with respect to prediction inaccuracies during the application of optimal control in real time.
- Application of the control system to a large-scale real combined sewer network that is located at the river Obere Iller in Bavaria is performed. The

evaluation of the efficiency of the methods is done on the basis of the results obtained for this network with different inflow scenarios. These investigations provided the basis for the development and implementation of the real control system actually operated in the sewer network of Obere Iller.

Chapter 4
Nonlinear Optimal Control

4.1 Performance Criterion

The main control objectives and the secondary operational objectives that were mentioned in Section 3.1 are considered *directly* in nonlinear optimal control (Marinaki and Papageorgiou, 1995). This is done via formulation of a nonlinear cost function that is minimized taking into account the state equation and the constraints.

The performance criterion to be minimized has the general form

$$J = \theta[\mathbf{x}(K)] + \sum_{k=0}^{K-1} \sum_{j=1}^{M} w_j \varphi_j[\mathbf{x}(k), \mathbf{u}(k), \mathbf{u}(k-1)] \tag{4.1}$$

where:
- K is the optimization time horizon.
- φ_j is the subgoal objective for $j = 1,...,M$.
- w_j is the weighting factor corresponding to subgoal φ_j.
- θ is the terminal cost function.

For the present problem, the performance criterion is suggested to contain the following five subgoals:
- For *avoiding overload* in storage elements without overflow, one may use

$$\varphi_1[\mathbf{x}(k)] = \sum_{i=1}^{n_y} \psi[V_{i,max} - V_i(k)]^2 \tag{4.2}$$

where $\psi[c] = \min[0, c]$ for $c \in R$, and n_y is the total number of storage elements of the sewer network without overflow capability.

- *Minimization of overflows* is pursued via

$$\varphi_2[\mathbf{x}(k)] = \sum_{i=1}^{n_x} S_i \, q_{over,i}^2(k) \tag{4.3}$$

where S_i are weighting factors and n_x is the total number of reservoirs of the sewer network that have overflow weirs. According to the choice of the weights S_i, the minimization of overflow from one reservoir can be regarded as more or less important than the minimization of overflow from other reservoirs.

- *Maximum utilization of the treatment plant* is pursued via

$$\varphi_3[\mathbf{u}(k)] = \sum_{i=1}^{n_r} [r_{i,max} - r_i(k)]^2 \tag{4.4}$$

where n_r is the total number of treatment plants. This term keeps the inflow to each treatment plant near its flow capacity so that the sewer network is emptied as soon as possible to provide free storage space for a possible future rainfall.

- Desired *distribution of reserve storage volume* is considered by use of

$$\varphi_4[\mathbf{x}(k)] = \sum_{i=1}^{n_x} [V_i(k) - v_i V_G(k)]^2 \tag{4.5}$$

where V_G is the sum of all reservoir storages at time k, that is,

$$V_G(k) = \sum_{i=1}^{n_x} V_i(k) \tag{4.6}$$

and v_i are chosen parameters such that

$$\sum_{i=1}^{n_x} v_i = 1, \quad 0 \le v_i = V_{i,max} / \sum_{i=1}^{n_x} V_{i,max} \le 1. \tag{4.7}$$

The subgoal φ_4 attempts to distribute the current global storage $V_G(k)$ according to the parameters v_i, that is, according to the reservoir storage capacities $V_{i,max}$. The desired distribution attempted via the subgoal φ_4 is meant to be active mainly during the emptying phase, so that equal free relative storages are provided to each reservoir in the event of a future rainfall.

- *Abrupt changes of outflows* are penalized via

$$\varphi_5[\mathbf{u}(k), \mathbf{u}(k-1)] = \sum_{i=1}^{nx} Q_i[u_i(k) - u_i(k-1)]^2 \tag{4.8}$$

in order to account for throttle gate inertia, where Q_i are weighting factors.

The terminal cost $\theta[\mathbf{x}(k)]$ is just a sum of the state-dependent subgoals φ_1, φ_2, and φ_4 for the final time K.

The use of quadratic terms in the formulation of the various subgoal objectives aims, besides minimization, at homogenizing in space and time the corresponding penalized quantities. For example, subgoal φ_2 in (4.3) attempts not only to minimize the overall overflows but also to distribute them to a certain extent in space and time, in accordance with the chosen weights S_i.

It should be noted that the criterion J represents a combination of different, partly competitive subgoals and can hardly be given an overall physical interpretation. It serves, by its definition, as a *vehicle* toward satisfactory control in the sense of a desired priority order between partially conflicting subgoals and restrictions. It is possible to specify a priority order for the subgoals by appropriate choice of the weighting factors w_j. The choice of the weighting factors w_j is performed via a preliminary *trial-and-error* procedure taking into account the desired priority order and the physical dimensions of the quantities involved. The trial-and-error procedure starts by giving some initial values to the weighting factors. Then, the nonlinear optimal control problem is solved for various representative inflow scenarios and, if the control results are not satisfactory, the values of the weighting factors are changed appropriately, the control problem is solved again and so forth. The desired priority order for the subgoals mentioned previously is $\varphi_1, \varphi_2, \varphi_3, \varphi_4, \varphi_5$, as the first three subgoals represent the main control objectives of the control system for sewer networks, and the last two subgoals are secondary operational objectives.

4.2 Mathematical Problem Formulation

The mathematical optimization problem to serve as a generic tool for central real-time combined sewer network control can be stated as follows:

For given inflows $\mathbf{d}(k)$, k=0,...K-1; given initial conditions $\mathbf{x}(0)$; given all retarded variables with negative time arguments; find the optimal trajectories $\mathbf{u}(k)$, $\mathbf{x}(k+1)$, k=0,...,K-1, minimizing the performance criterion (4.1) subject to the state Equation (2.37) and the constraints (2.38).

4.3 Solution Algorithm

4.3.1 General Problem Formulation

For the solution of the nonlinear optimal control problem a *feasible direction algorithm* is used that has been developed applying the discrete maximum principle (Papageorgiou, 1996). In its present form, the algorithm has been extended to consider directly the state-dependent control constraints (4.12), which improves significantly the computational efficiency (Papageorgiou and Marinaki, 1995).

The *general* problem considered in the solution algorithm is that of minimizing a cost function

$$J = \theta[\mathbf{x}(K)] + \sum_{k=0}^{K-1} \varphi[\mathbf{x}(k), \mathbf{u}(k), k] \tag{4.9}$$

subject to the state equation

$$\mathbf{x}(k+1) = \mathbf{f}[\mathbf{x}(k), \mathbf{u}(k), k], \quad k = 0, \ldots, K-1 \tag{4.10}$$

$$\mathbf{x}(0) = \mathbf{x}_0 \tag{4.11}$$

and the constraints

$$\mathbf{u}_{min}[\mathbf{x}(k), k] \le \mathbf{u}(k) \le \mathbf{u}_{max}[\mathbf{x}(k), k], \quad k = 0, \ldots, K-1 \tag{4.12}$$

that may be brought to the general form

$$\mathbf{h}[\mathbf{x}(k), \mathbf{u}(k), k] \le \mathbf{0}, \quad k = 0, \ldots, K-1 \tag{4.13}$$

where θ, φ, $\mathbf{h} \in R^q$ are twice continuously differentiable functions, and $\mathbf{x} \in R^n$ and $\mathbf{u} \in R^m$ are the state and control vectors, respectively.

4.3.2 Necessary Optimality Conditions

The extended discrete-time Hamiltonian function for this problem is

$$\tilde{H}[\mathbf{x}(k), \mathbf{u}(k), \lambda(k+1), \mu(k), k] = \varphi[\mathbf{x}(k), \mathbf{u}(k), k] + \lambda(k+1)^T \mathbf{f}[\mathbf{x}(k), \mathbf{u}(k), k] +$$
$$+ \mu(k)^T \mathbf{h}[\mathbf{x}(k), \mathbf{u}(k), k] \tag{4.14}$$

where $\lambda(k+1) \in R^n$ and $\mu(k) \in R^q$ are the *Lagrange* and *Kuhn-Tucker* multipliers, respectively, for the corresponding equality and inequality conditions. The gradient of J with respect of **u**, taking into account the equality constraints (4.10), is given by (notation: $x_y = \partial x / \partial y$)

$$g(k) = \varphi_{\mathbf{u}(k)} + \mathbf{f}_{\mathbf{u}(k)}^T \lambda(k+1) \tag{4.15}$$

where the variables $\lambda \in R^n$ satisfy

$$\lambda(k) = \widetilde{H}_{\mathbf{x}(k)}, \quad k=0,...,K-1 \tag{4.16}$$

$$\lambda(K) = \theta_{\mathbf{x}(K)}. \tag{4.17}$$

The reduced gradient γ, with regard to the inequality constraints (4.12), has the components

$$\gamma_i(k) = \begin{cases} 0 & \text{if } u_i(k) = u_{i,min}[x_i(k),k] \text{ and } p_i(k) < 0 \\ 0 & \text{if } u_i(k) = u_{i,max}[x_i(k),k] \text{ and } p_i(k) > 0 \\ g_i(k) & \text{else} \end{cases} \tag{4.18}$$

where $p_i(k)$ is a search direction. The scalar product of two vector trajectories $\mathbf{\eta}(k)$, $\xi(k)$, for $k=0,...,K-1$, is defined as follows:

$$[\mathbf{\eta}(K), \xi(K)] = \sum_{k=0}^{K-1} \mathbf{\eta}(k)^T \xi(k). \tag{4.19}$$

Furthermore, a saturation vector function $\mathbf{sat}(\mathbf{\eta})$ is defined to have the components

$$\text{sat}_i(\mathbf{\eta}) = \begin{cases} \eta_{i,max} & \text{if } \eta_i > \eta_{i,max} \\ \eta_{i,min} & \text{if } \eta_i < \eta_{i,min} \\ \eta_i & \text{else.} \end{cases} \tag{4.20}$$

With the above definitions, the necessary conditions for optimality are given by (4.10), (4.11), (4.13), (4.16), (4.17), and

$$\widetilde{H}_{\mathbf{u}(k)} = \mathbf{0} \tag{4.21}$$

$$\mu(k)^T \mathbf{h}[\mathbf{x}(k),\mathbf{u}(k),k] = 0 \tag{4.22}$$

$$\mu(k) \geq \mathbf{0}. \tag{4.23}$$

Hence, if these equations are satisfied simultaneously by some trajectories $\mathbf{x}(k+1)$, $\mathbf{u}(k)$, $\lambda(k)$, and $\mu(k)$, a stationary point of the optimal control problem has been found. Defining $h_i[\mathbf{x}(k),\mathbf{u}(k),k]=[u_i(k)-u_{i,min}(x_i(k),k)][u_i(k)-u_{i,max}(x_i(k),k)]$, it follows from (4.15) and (4.21)

$$
\mu_i(k) = \begin{cases} -g_i(k)/[2u_i(k) - u_{i,min}[x_i(k), k] - u_{i,max}[x_i(k), k]] \\ \qquad \text{if } u_i(k) = u_{i,min}[x_i(k), k] \text{ or } u_i(k) = u_{i,max}[x_i(k), k] \\ 0 \qquad\qquad\qquad \text{else.} \end{cases} \qquad (4.24)
$$

4.3.3 Structure of the Solution Algorithm

The necessary conditions for optimality constitute a two-point boundary value problem (TPBVP). A solution algorithm for this problem can be obtained by using the following feasible direction algorithm:

Step 1: Select a feasible initial control trajectory $\mathbf{u}^0(k)$, k=0,...,K–1. Set the iteration index i=0.

Step 2: Using $\mathbf{u}^i(k)$, k=0,...,K–1, solve (4.10) from the initial condition (4.11) to obtain $\mathbf{x}^i(k+1)$.

Step 3: Using $\mathbf{u}^i(k)$, $\mathbf{x}^i(k+1)$, k=0,...,K–1, solve (4.16) from the terminal condition (4.17) to obtain $\lambda^i(k)$ and calculate the gradients $\mathbf{g}^i(k)$ and $\gamma^i(k)$ from (4.15) and (4.18), respectively, and Kuhn-Tucker multipliers $\mu^i(k)$ from (4.24).

Step 4: Specify a search direction $\mathbf{p}^i(k)$, k=0,...,K–1.

Step 5: Apply a one-dimensional search routine along the $\mathbf{p}^i$-direction to obtain $\mathbf{u}^{i+1}(k)$. The corresponding line-optimization problem reads:

$$
\min_{\alpha>0} J\{\mathbf{sat}[\mathbf{u}^i(k) + \alpha\mathbf{p}^i(k)]\}
$$

where α is a scalar step length. (During the solution of the line optimization problem, repeated calculations of the states $\mathbf{x}(k+1)$, the gradients $\mathbf{g}(k)$ and $\gamma(k)$, and the multipliers $\lambda(k)$ and $\mu(k)$ are performed.) Then, calculate the controls from $\mathbf{u}^{i+1}(k) = \mathbf{sat}[\mathbf{u}^i(k) + \alpha^i\mathbf{p}^i(k)]$.

Step 6: If for a given scalar $\varepsilon>0$ the inequalities

$$[\gamma^i(K), \gamma^i(K)] < \varepsilon \quad \text{and} \quad \mu^i(k)\geq0, \quad k=0, ...,K-1$$

hold, stop. Otherwise, set i=i+1 and go to step 2. [All necessary optimality conditions are satisfied by the algorithm if $\gamma(k) = \mathbf{0}$ and $\mu(k) \geq \mathbf{0}$ $\forall k\in[0, K-1]$ (Papageorgiou and Marinaki, 1995).]

The necessary conditions and the solution algorithm may be easily extended to consider variables with time delays such as those appearing in Equations (2.37), (2.38b), and (4.1) (Papageorgiou and Marinaki, 1995).

In cases of a convex objective function (4.9) and linear model Equations (4.10) and constraints (4.12), a global minimum will be reached at convergence. In the general nonlinear case, the algorithm will reach at least a local minimum at convergence. No difficulties with unsatisfactory local minima were encountered when solving the sewer network control problem for different inflow scenarios.

It is important to note that in this algorithm at the beginning of each iteration, the active bounds, for which $\gamma_i(k) = 0$ according to (4.18), are memorized by the algorithm. These active bounds are forced to remain active until the end of the corresponding iteration; that is, the concerned controls are not calculated from $\text{sat}_i[u^i(k) + \alpha p^i(k)]$ but from $u_{i,max}(x(k),k)$ or $u_{i,min}(x(k),k)$, according to which bound was activated. As a consequence, the projected gradient components $\gamma_i(k)$ of the memorized actively bounded controls are not calculated from (4.18) but are set equal to zero for the duration of the corresponding iteration. This measure is necessary to guarantee consistency of the respective calculations of $\gamma_i(k)$ and $u_i(k)$ during the iteration. To see this, consider for example the case where $u_i(k) = u_{i,max}(x(k),k)$ and $p_i(k){>}0$ at the beginning of the iteration; then $\gamma_i(k) = 0$ according to (4.18), that is, $u_i(k)$ is expected to remain on its upper bound for positive line steps α. However, because $u_{i,max}$ depends on $x(k)$, and $x(k)$ depends on all previous $u(\kappa)$, $\kappa{=}0,...,$ $k{-}1$, a line step α may increase $u_{i,max}(x(k),k)$ more rapidly than $u_i(k)$, that is, deactivate the control bound. In this case, $\gamma_i(k) = 0$ is incorrect and leads to an erroneous calculation of $J'(\alpha)$. These situations are avoided, and consistency of calculations is established, if, as mentioned above, $u_i(k)$ is forced to remain equal to $u_{i,max}(x(k),k)$ for the duration of the corresponding iteration, in which case $\gamma_i(k) = 0$ is correct. In some practical applications, it was found that this measure decelerates to some extent the algorithm's convergence. To avoid this, the measure may be applied only after the manifest appearance of line search difficulties, that is, when for two continuous iterations of the algorithm a restart is effectuated due to problems in the sectioning or bracketing phases as those mentioned in Section 4.3.6, due to the mentioned inconsistency.

4.3.4 Specification of a Search Direction

Several methods can be used in step 4 of the algorithm for the specification of the search direction p^i (Bunday, 1984; Scales, 1985; Fletcher, 1987). All these methods use a search direction that satisfies $(p^i, g^i) < 0$, which guarantees that the derivative $dJ/d\alpha$ is always negative for $\alpha{=}0$ (except if u^i is a stationary point) and therefore the objective function can be improved for some $\alpha^i {>}0$ (Gill et al., 1981). The control constraints are taken into account by using the reduced gradient γ^i for all methods.

Some of these techniques are the steepest descent method, the conjugate gradient methods, and the quasi-Newton methods.

Steepest Descent

Steepest descent is the simplest but unfortunately the least efficient method for the specification of a search direction. It is based on the steepest descent direction, namely (for convenience the time index k is omitted, but each of the formulas below is executed for all k=0,...,K−1)

$$p^i = -g^i. \tag{4.25}$$

Steepest descent typically leads to a quick approach of the minimum, if the initial guess $\mathbf{u}^0$ was chosen far from the minimum. However, in a vicinity of the minimum the method is known to be slow.

Conjugate Gradient Methods

Conjugate gradient methods approximate the simplicity of steepest descent and the efficiency of quasi-Newton methods. There are two main conjugate gradient methods, namely *Fletcher-Reeves (FR)* and *Polak-Ribiere (PR)*. The search direction according to Fletcher-Reeves is as follows:

$$\mathbf{p}^i = -\mathbf{g}^i + \beta^i \mathbf{p}^{i-1} \tag{4.26}$$

where $\beta^0 = 0$, and for $i \geq 1$

$$\beta^i = (\gamma^i, \gamma^i)/(\gamma^{i-1}, \gamma^{i-1}). \tag{4.27}$$

For $i = 0$ we have $\mathbf{p}^0 = -\mathbf{g}^0$, that is, the method starts with the steepest descent direction. For Polak-Ribiere we also have (4.26) but β^i is calculated from

$$\beta^i = ((\gamma^i - \gamma^{i-1}), \gamma^i)/(\gamma^{i-1}, \gamma^{i-1}). \tag{4.28}$$

The Fletcher-Reeves method requires the storage of three, whereas the Polak-Ribiere method requires the storage of four mK-dimensional arrays, thus these methods can easily be applied to problems with hundreds or even thousands of variables. Both methods produce descent directions if the line search is exact, as can be easily shown.

Quasi-Newton Methods

Quasi-Newton methods require more complex calculations for the specification of a search direction. There are two main quasi-Newton methods, the *DFP* proposed in 1963 by Davidon, Fletcher, and Powell, and the *BFGS* proposed independently in 1970 by Broyden, Fletcher, Goldfarb, and Shanno.

The search direction calculation according to DFP is as follows:

$$\mathbf{p}^0 = -\mathbf{g}^0 \tag{4.29}$$

and for $i \geq 1$

$$\sigma^{i-1} = \mathbf{u}^i - \mathbf{u}^{i-1} \tag{4.30}$$

$$\mathbf{y}^{i-1} = \gamma^i - \gamma^{i-1} \tag{4.31}$$

$$\mathbf{z}^{i-1} = \mathbf{y}^{i-1} + \sum_{j=0}^{i-2} [(\mathbf{v}^j, \mathbf{y}^{i-1})\mathbf{v}^j - (\mathbf{w}^j, \mathbf{y}^{i-1})\mathbf{w}^j] \tag{4.32}$$

$$\mathbf{v}^{i-1} = (\ \mathbf{\sigma}^{i-1}, \mathbf{y}^{i-1})^{-1/2} \mathbf{\sigma}^{i-1}$$ (4.33)

$$\mathbf{w}^{i-1} = (\ \mathbf{z}^{i-1}, \mathbf{y}^{i-1})^{-1/2} \mathbf{z}^{i-1}$$ (4.34)

$$\mathbf{p}^i = -\mathbf{g}^i - \sum_{j=0}^{i-1} [(\mathbf{v}^j, \mathbf{\gamma}^i)\mathbf{v}^j - (\mathbf{w}^j, \mathbf{\gamma}^i)\ \mathbf{w}^j].$$ (4.35)

Because of the sums in (4.32) and (4.35), the vectors **v**, **w** of *all* previous iterations j=0,...,i−1 must be stored. This means that the storage requirements of the method increase steadily during application. This increase may be limited, however, through periodic restart (see Section 4.3.6).

The search direction calculation according to BFGS is as follows:

$$\mathbf{p}^0 = -\mathbf{g}^0$$ (4.36)

and for i ≥1 and for $\mathbf{\sigma}^{i-1}$, $\mathbf{y}^{i-1}$, $\mathbf{v}^{i-1}$ and $\mathbf{w}^{i-1}$ as in Equations (4.30), (4.31), (4.33), and (4.34), respectively, of the DFP method:

$$\mathbf{z}^{i-1} = \mathbf{y}^{i-1} + \sum_{j=0}^{i-2} [(\mathbf{v}^j, \mathbf{y}^{i-1})\mathbf{v}^j - (\mathbf{w}^j, \mathbf{y}^{i-1})\mathbf{w}^j + (\mathbf{b}^j, \mathbf{y}^{i-1})\mathbf{b}^j]$$ (4.37)

$$\mathbf{b}^{i-1} = (\ \mathbf{\sigma}^{i-1}, \mathbf{y}^{i-1})^{-1/2} (\mathbf{y}^{i-1}, \mathbf{z}^{i-1})^{1/2} \mathbf{v}^{i-1} - \mathbf{w}^{i-1}$$ (4.38)

$$\mathbf{p}^i = -\mathbf{g}^i - \sum_{j=0}^{i-1} [(\mathbf{v}^j, \mathbf{\gamma}^i)\mathbf{v}^j - (\mathbf{w}^j, \mathbf{\gamma}^i)\ \mathbf{w}^j + (\mathbf{b}^j, \mathbf{\gamma}^i)\ \mathbf{b}^j].$$ (4.39)

These formulas indicate that the computational effort per iteration for BFGS is higher than for DFP, as BFGS requires storage of three vectors (**v**, **w**, **b**) from all previous iterations j = 1,...,i−1, whereas DFP requires storage only of two vectors (**v**, **w**).

Both BFGS and DFP may be shown to produce descent directions of search. For exact line search, BFGS and DFP produce identical results, but for inexact line search they may show different efficiency.

4.3.5 Line Search Algorithm

In step 5 of the solution algorithm there is the subproblem of finding the line minimum on a specified search direction that uses calculations of $J\{\mathbf{sat}[\mathbf{u}^i(k) + \alpha \mathbf{p}^i(k)]\}$ for corresponding values of α. The line minimum satisfies $dJ/d\alpha = 0$.

The line search algorithm specifies in corresponding iterations a sequence of steps $\{\alpha_j\}$ and terminates when a step has been found that satisfies the following two conditions (the superscript i is omitted for convenience):

$$J(\alpha) \le J(0) + \alpha \rho\ J'(0)$$ (4.40)

$$|J'(\alpha)| \le -\sigma\ J'(0)$$ (4.41)

where $\rho \in [0, 1/2]$ and $\rho \leq \sigma < 1$ are pre-chosen parameters and $J(\alpha) = J\{\mathbf{sat}[\mathbf{u}^i(k) + \alpha \mathbf{p}^i(k)]\}$.

Condition (4.40) provides an upper bound for the line search and guarantees that there is improvement of the cost function at each iteration, whereas condition (4.41) guarantees that the improvement of the cost function is not negligible. The accuracy of line search is higher if σ is closer to zero. A highly accurate line search may require many iterations and reduce the efficiency of the overall optimal control algorithm. A low-accuracy line search may also reduce the efficiency of the overall algorithm leading to a higher number of overall iterations. An appropriate choice of σ depends on the concrete application problem but also on the employed search direction method.

The line search algorithm includes the *bracketing phase* that attempts to locate a bracket $[a_i, b_i]$ (the bracket is characterized by $J'(a_i) < 0$ and $J'(b_i) > 0$) that includes the line minimum, and the *sectioning phase* that produces a sequence of brackets $[a_j, b_j]$ with diminishing length. The sectioning phase involves polynomial interpolation.

Bracketing Phase

The bracketing phase starts with $\alpha_0 = 0$ and an initial step α_1, for example $\alpha_1 = -2(J^{i-1} - J^i)/J'(0)$. This choice of α_1 is found using quadratic interpolation. Each iteration i of the bracketing phase ends up either with specification of a line minimum (in which case no sectioning phase takes place), or with specification of a bracket to be passed over to the sectioning phase, or with a new step α_{i+1} to be passed over to the next bracketing iteration. To this end, each bracketing iteration executes the following steps (Fletcher, 1987; Papageorgiou and Marinaki, 1995):

Step 1: Calculation of $J(\alpha_i)$.

Step 2: If $J(\alpha_i) > J(0) + \alpha_i \rho J'(0)$ or $J(\alpha_i) \geq J(\alpha_{i-1})$, then
$a_i = a_{i-1}$, $b_i = \alpha_i$ gives a bracket that includes the line minimum and the bracketing phase is terminated.

Step 3: Calculation of $J'(\alpha_i)$.

Step 4: If $|J'(\alpha_i)| \leq -\sigma J'(0)$, the line search is terminated.

Step 5: If $J'(\alpha_i) \geq 0$ then
$a_i = a_{i-1}$, $b_i = \alpha_i$ gives a bracket that includes the line minimum and the bracketing phase is terminated
else
set $\alpha_{i+1} = \alpha_i + \tau_1 (\alpha_i - \alpha_{i-1})$, where $\tau_1 \geq 1$. (For example, with $\tau_1 = 2$ the step is doubled at each iteration of the bracketing phase. This speeds up the bracketing phase if the initial step α_1 was chosen too short.)

Sectioning Phase

The sectioning phase also executes a series of iterations j, each ending either with a step α_j that satisfies (4.40) and (4.41), in which case the line search terminates, or with a new shorter bracket $[a_j, \alpha_j]$ or $[\alpha_j, b_j]$, for the next iteration. Each sectioning iteration j executes the following steps:

Step 1:   Choice of $\alpha_j \in [a_j + \tau_2(b_j-a_j), b_j - \tau_3(b_j - a_j)]$ via interpolation (quadratic or cubic interpolation).

Step 2:   Calculation of $J(\alpha_j)$.

Step 3:   If $J(\alpha_j) > J(0) + \rho\alpha_j J'(0)$ or $J(\alpha_j) \geq J(a_j)$, then
$a_{j+1} = a_j$, $b_{j+1} = \alpha_j$ is a shorter bracket for the next iteration.

Step 4:   Calculation of $J'(\alpha_j)$.

Step 5:   If $|J'(\alpha_j)| \leq -\sigma J'(0)$, the line search is terminated.

Step 6:   If $J'(\alpha_j) \geq 0$ then
$a_{j+1} = a_j$, $b_{j+1} = \alpha_j$ is a shorter bracket for the next iteration
else
$a_{j+1} = \alpha_j$, $b_{j+1} = b_j$ is a shorter bracket for the next iteration.

The predefined parameters τ_2, τ_3, $0 < \tau_2 < \tau_3 \leq 0.5$ guarantee that there will be a nonnegligible reduction of the bracket from iteration to iteration and it then follows that

$$|b_{j+1} - a_{j+1}| \leq (1-\tau_2)|b_j - a_j| \qquad (4.42)$$

which guarantees the convergence of the algorithm. The step α_j that is found at convergence is the step α^i of the corresponding overall iteration.

4.3.6 Restart

Due to nonquadratic cost functions or due to inexact line search, the algorithm may encounter difficulties during the iterations. A degenerated search direction may be detected by being approximately orthogonal to the gradient, or even by being an ascent direction. For this reason, the following *condition of sufficient negativity* should be checked at each iteration:

$$(\mathbf{p}^i, \gamma^i) \leq -B[(\mathbf{p}^i, \mathbf{p}^i)(\gamma^i, \gamma^i)]^{1/2} \qquad (4.43)$$

where B is a suitable positive parameter. If (4.43) is violated, then a *restart* should be effectuated. Restart means to interrupt the running iteration and start a new one with steepest descent. The following iterations may then continue normally, according to the chosen search direction method.

Independently of (4.43), a *periodic restart* every N iterations, where N<<mK, is known to improve the efficiency of the algorithm. Moreover, in case of application of a quasi-Newton method, a periodic restart limits the amount of required computer space and the computing time per iteration.

Moreover, restart is effectuated during a bracketing iteration i if for a step α_i the value $J'(\alpha_i)$ is less than zero while the value $J(\alpha_i)$ is greater than $J(\alpha_{i-1})$ (which may occur if the function to be minimized is non-convex) or during the sectioning phase when the number of iterations exceeds a prespecified number while the equations (4.40) and (4.41) are not yet satisfied.

4.3.7 Algorithm Comparisons

In the feasible direction algorithm presented previously, an important observation concerns the constraints (2.38b). These constraints can be regarded either indirectly, by introducing penalty terms in the objective function, or directly by using the *projection method* (Rosen, 1960, Kirk, 1970) in the feasible direction algorithm. However, it should be noted that the modifications needed to incorporate the projection procedure in the solution algorithm (Papageorgiou and Marinaki, 1995) increase the code significantly. Moreover, the projection procedure increases the computational effort per iteration. Thus, in practical applications, it should be checked if the indirect approach with the penalty terms is beneficial with respect to the computational effort.

For the sewer network control problem that is studied here, both methods, each with or without the measure concerning the x-dependent active control bounds (see Section 4.3.3), were tested initially for an optimization horizon equal to 6 h. More precisely, the comparison of the following *four approaches* was performed:

- The *first approach* uses penalty terms in the cost function to take into account the control constraints (2.38b) and forces x-dependent control bounds, which are active at the beginning of an iteration, to remain active until the end of the corresponding iteration (curve k1 in Figure 4.1).
- The *second approach* uses the same method of the first approach regarding the control constraints, but the measure concerning the active control bounds is automatically switched on only when line search difficulties appear, which usually may be the case only when the algorithm has approached the minimum (curve k2 in Figure 4.1).
- The *third approach* uses the projection method to take into account the control constraints (2.38b) and the measure concerning the active control bounds for all iterations (curve k3 in Figure 4.1).
- The *fourth approach* uses again the projection method, but the measure concerning the active control bounds is applied only in the final iterations (curve k4 in Figure 4.1).

It may be seen from Figure 4.1 that when the control constraints are considered indirectly by introducing penalty terms in the objective function (curves k1, k2), the computation time needed for reaching the minimum is less than in the corresponding opposite case (curves k3, k4). More precisely, the CPU times in Workstation HP 700 required for practical convergence of the algorithm are 3.7 min, 2.8 min, 6.1 min, and 7.2 min for the four respective approaches.

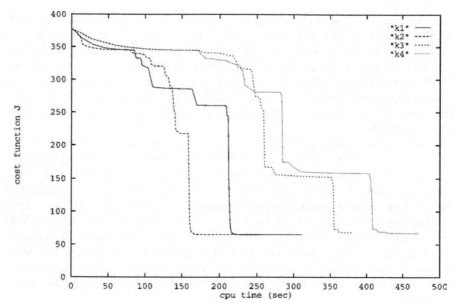

Figure 4.1. Decreasing cost function values in dependence of the computing time for the search direction method DFP using the four approaches.

These results indicate that the second approach needs the least computation time for reaching the minimum (2.8 min) and is the most efficient method for the numerical solution of the present problem. For this reason we apply in the following discussion this approach for the calculation of the optimal state and control trajectories.

Using this approach a comparison between the methods used for the specification of the search direction, namely the steepest descent method, the quasi-Newton methods DFP and BFGS, and the conjugate gradient methods of Fletcher-Reeves and Polak-Ribiere, is effectuated so as to find their respective efficiencies with respect to the computation time needed for numerical solution of the particular sewer network control problem. Scaling was not found to improve the computational efficiency of any method.

The search direction methods differ from each other with respect to the mean computation time they require for each iteration, and with respect to the mean performance function amelioration they produce at each iteration. A clear picture of the performance of the different methods is provided in Figure 4.2, which depicts the objective function value versus the required CPU time (in a Workstation HP 700) for each method after 500 iterations.

The two numbers appearing in parenthesis next to each method, refer to the frequency of restarts (see Section 4.3.6) and the value of σ (see Section 4.3.5), respectively, in the feasible direction algorithm. These two parameters have an impact on each method's efficiency, and they were given the indicated values after a preliminary investigation aiming at maximising the performance of the concerned method.

The quasi-Newton and conjugate gradient methods perform a significant amelioration of the objective function within few minutes. It can be seen (Figure 4.2, Table 4.1), for example, that DFP (20, 0.5) reduces the value of the cost function virtually to its minimum in 2.8 min. Afterward, there is virtually no amelioration of the objective function. Taking into account the computation time needed and the amelioration of the cost function, we can say that DFP (20, 0.5) is the most efficient method for the numerical solution of the particular optimal control problem.

The achieved computation times ensure the real-time applicability of the method. It should be noted that in the real-time applications the optimization horizon is less than 6 h (see Section 7.2.3), which leads to accordingly less computation time for virtually reaching the cost function minimum. It should also be noted that if the algorithm encounters convergence difficulties leading to long computation times, it may be interrupted (e.g., after a maximum time is reached) with the best solution reached until this iteration. This is because the algorithm is feasible, that is, all constraints are satisfied at any iteration. Finally, it should be emphasized that modern PCs are 20 times faster than the utilized workstation, hence computation time is not real-time critical even for large-scale sewer networks.

It should be noted that the solution algorithm may reach a local minimum in the general nonlinear case. However, for the sewer network control problem, using different network topologies, different inflow scenarios, and different initial guess trajectories, we always encountered a highly efficient solution. Therefore, it appears that local minima do not present a real difficulty for the problem at hand.

4.3.8 RPROP Algorithm

Except for the feasible direction algorithm (Section 4.3.3), the *RPROP algorithm* (*R*esilient Back*prop*agation) (Riedmiller and Braun, 1993; Riedmiller, 1994) has also been tested for the solution of the nonlinear optimal control problem. The RPROP algorithm is an adaptive technique that does not consider the value of the gradient $\mathbf{g}$ but is only dependent on the sign of its components. This algorithm does not calculate a search direction nor does it solve the line optimization problem (steps 4 and 5, respectively, of the solution algorithm of Section 4.3.3) but instead it changes directly the correction $\mathbf{d}^i$ that is used in the calculation of the new control variable:

$$\mathbf{u}^{i+1}(k)=\mathbf{sat}[\mathbf{u}^i(k)+\mathbf{d}^i(k)]. \tag{4.44}$$

Table 4.1. Computation time needed for virtually reaching the cost function minimum.

Method	Steepest Descent (0.2)	Fletcher-Reeves (240, 0.5)	Polak-Ribiere (240, 0.2)	DFP (20, 0.5)	BFGS (20, 0.8)
Comp. Time (min)	4.7	3.3	3.3	2.8	4.3

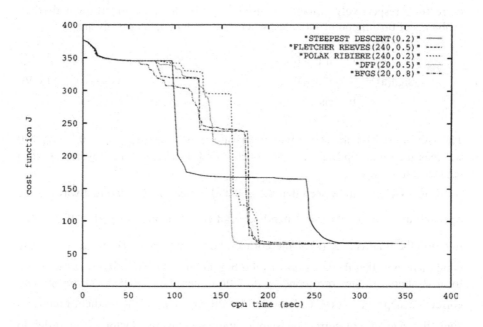

Figure 4.2. Decreasing cost function values in dependence of the computing time.

More precisely, the iterative RPROP algorithm has the following form:

Step 1: Select a feasible initial control trajectory $\mathbf{u}^0(k)$ and an initial descent-direction correction $\mathbf{d}^{-1}(k)$, k=0,...,K–1. Set the iteration index i=0.

Step 2: Using $\mathbf{u}^i(k)$, k=0,...,K–1, solve (4.10) from the initial condition (4.11) to obtain $\mathbf{x}^i(k+1)$.

Step 3: Using $\mathbf{u}^i(k)$, $\mathbf{x}^i(k+1)$, k=0,...,K–1, solve (4.16) from the terminal condition (4.17) to obtain $\lambda^i(k)$ and calculate the gradients $\mathbf{g}^i(k)$ and $\gamma^i(k)$ from (4.15) and (4.18), respectively, and Kuhn-Tucker multipliers $\mu^i(k)$ from (4.24). [For the calculation of $\gamma^i(k)$ from (4.18), the correction $\mathbf{d}^{\,i-1}(k)$ is regarded as the search direction $\mathbf{p}^i(k)$.]

Step 4: Specify the correction $\mathbf{d}^i(k)$, k=0,...,K–1 (see below).

Step 5: Calculate the updated controls from (4.44).

Step 6: If for a given scalar ε>0 the inequalities

$$[\gamma^i(K), \gamma^i(K)] < \varepsilon \quad \text{and} \quad \mu^i(k) \geq 0, \quad k=0, ..., K-1$$

hold, stop. Otherwise, set i=i+1 and go to step 2.

The updated values $d_j^i(k)$ are given by

$$d_j^i(k) = \begin{cases} -\text{sign}(g_j^i(k)) \cdot \min\{\eta^+ \mid d_j^{i-1}(k) \mid, d_{max}\} & \text{if } (g_j^{i-1}(k) \cdot g_j^i(k)) > 0 \\ -\text{sign}(g_j^i(k)) \cdot \max\{\eta^- \mid d_j^{i-1}(k) \mid, d_{min}\} & \text{if } (g_j^{i-1}(k) \cdot g_j^i(k)) < 0 \end{cases} \quad (4.45)$$

where $0 < \eta^- < 1 < \eta^+$, and d_{min} and d_{max} are the lower and the upper limits of the correction, respectively, which are used in order to avoid overflow/underflow problems of floating point variables, whereas the sign operator is given by

$$\text{sign}(a) = \begin{cases} 1 & \text{if } a > 0 \\ -1 & \text{if } a < 0 \\ 0 & \text{else.} \end{cases} \quad (4.46)$$

The values used for the above parameters are $\eta^-=0.5$, $\eta^+=1.2$, $d_{min}=10^{-6}$, $d_{max}=50$ whereas the initial update trajectory is $d^{-1}(k)=0.1$. These choices were found to lead to best results.

From (4.45) it can be seen that the updated corrections $d_j^i(k)$ are always in the opposite direction of the corresponding gradient components $g_j^i(k)$. If at some iteration the gradient $g_j^i(k)$ of the corresponding control $u_j^i(k)$ changes its sign, which indicates that the last step was too big and the algorithm has jumped over a local minimum in the corresponding direction, then the size of the corresponding correction step $\mid d_j^{i-1}(k) \mid$ is decreased by the factor η^-; if the gradient retains its sign, the size of the correction step is increased by the factor η^+ in order to accelerate convergence. It should be noted that the RPROP algorithm does not guarantee a decrease of the cost function value at each iteration.

Figure 4.3 illustrates the results obtained when applying the feasible direction algorithm and the RPROP algorithm to the sewer network control problem that is studied in this book. It should be noted that, in order to safely and accurately locate the minimum when applying the RPROP procedure, an automatic switch to the feasible direction algorithm is performed when the cost function value is not decreased for a pre-specified number of successive iterations. It can be seen that the RPROP algorithm reaches the minimum in less computation time than the feasible direction algorithm. These results indicate that the PRPOP algorithm is an efficient method for the numerical solution of the sewer network control problem. However, as the analysis of the PRPOP algorithm started at a late phase of this study, the feasible direction algorithm is used in the following sections for the calculation of the optimal state and control trajectories.

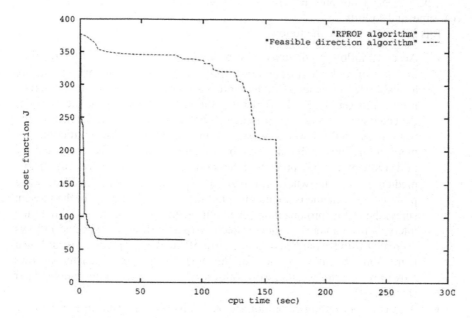

Figure 4.3. Comparison between the RPROP and the feasible direction algorithm.

4.4 Rolling Horizon

Due to the simplifications in the model of the sewer network that is used for nonlinear optimal control and the inaccuracies of the predictions of external inflows, which are usually sufficiently accurate only for a relatively short future period, the optimization should better be repeated periodically, every k_R time steps, with updated inflow predictions and state estimates. Thus, the optimal control problem of the sewer network is embedded in a closed-loop control structure (referred as *optimization with rolling horizon*), where the numerical solution in real time is performed as follows (Papageorgiou, 1997):

At time step k_0, the nonlinear optimal control problem is solved (using the iterative feasible direction algorithm), based on the estimated initial condition $\mathbf{x}(k_0)$ and on the available predictions of external inflows $\mathbf{d}(k)$, $k=k_0,\ldots,k_0+K-1$, where K ($K \gg k_R$) is the optimization horizon, to obtain the trajectories of the controls $\mathbf{u}^*(k)$ and states $\mathbf{x}^*(k+1)$, for $k=k_0,\ldots,k_0+K-1$. However, only a part of the control trajectory is actually applied to the process, namely $\mathbf{u}^*(k)$, $k=k_0,\ldots,k_0+k_R-1$. Then, at time step k_0+k_R, based on new estimated initial condition $\mathbf{x}(k_0+k_R)$ and updated predictions of external inflows $\mathbf{d}(k)$, $k=k_0+k_R,\ldots,k_0+k_R+K-1$, the optimization problem is solved again to obtain the trajectories for $\mathbf{u}^*(k)$ and $\mathbf{x}^*(k+1)$ for $k=k_0+$

$k_R,\ldots,k_0+ k_R+K-1$ but only $\mathbf{u}^*(k)$ for $k=k_0+k_R,\ldots,k_0+2k_R-1$, is actually applied in real time, and so forth.

It is important to note the following:

- Accurate inflow predictions may be available only for the first K_p time intervals of each optimization run, where $K_p \leq K$ (*e.g.*, if no predictions are available, $K_p=0$), whereas in the (unrealistic) case of accurate and complete inflow information, $K_p=K$. Generally, the K_p value depends on the extent of the catchment area (*e.g.*, in a large catchment area, the time needed for the water to reach the sewer network is longer, and thus inflow predictions for more future time intervals can be obtained) and can be increased by utilization of rainfall prediction methods or radars. However, as inflow predictions for the whole optimization horizon $K \geq K_p$ are needed, some prolongation scheme is applied for the inflows after time K_p. In the present study, the inflow prolongation scheme uses the values of the last three time intervals from a known inflow trajectory (past inflows or expected inflows estimated using a predictive rainfall-runoff model) to predict, using linear regression, the inflow values for the next 20 minutes. Then the inflows move towards dry weather flow values, which they reach 20 minutes later (Figure 4.4).
- A satisfactory optimization horizon K is given by the sum of the duration of the predicted inflow and the time needed for the network to be emptied (Papageorgiou, 1988). A much smaller optimization horizon K may lead to "myopic" control actions.
- For the application of the repeated optimization, the computation time needed for the numerical solution of the problem must be short enough to permit repetitive on-line solution of the optimization problem.
- The state variables $\mathbf{x}$ must be measurable or be estimated in real time. For the sewer network control problem, measurement sensors are used to obtain current water flow or water level measurements, mainly for the reservoirs and possibly for some link outflows. If for some elements of the network real-time measurements are lacking, a state observer must be employed to estimate the missing measurements in real time (see Section 5.3).

Generally, we can say that the rolling horizon method represents a closed-loop structure as the control decisions are taken every k_R based on measured or estimated states, and in this way the influence of inaccurate predictions, modelling inaccuracies, and diverse disturbances is reduced.

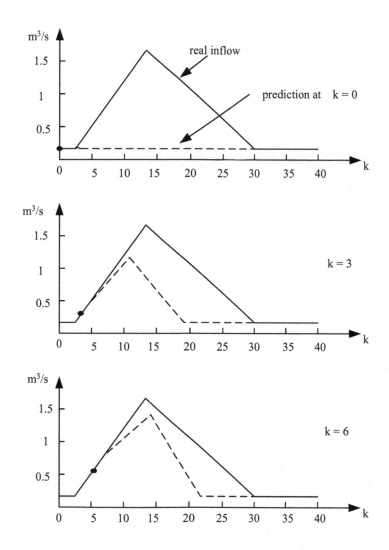

Figure 4.4. Example of inflow prediction at k = 0, 3, 6 for $k_R = K_p = 3$.

Chapter 5
Multivariable Feedback Control

5.1 General Problem Considerations

For the development of the linear multivariable feedback regulator, theoretically founded linear control theory methods, such as pole assignment or linear-quadratic (LQ) optimization, may be used. Application of these methods requires a number of problem simplifications, such as model linearization, quadratic criterion, and no constraints, which will be introduced in this chapter. For the sewer network flow control problem, application of the linear-quadratic methodology appears most convenient (Messmer and Papageorgiou, 1992). The linear-quadratic design procedure includes precise specifications on model structure, model equations, nominal steady-state choice, and quadratic criterion choice. These specifications enable system designers to come up in a short time with a linear-quadratic formulation of their particular sewer control problem and subsequently to proceed to a systematic and straightforward calculation of a multivariable feedback regulator based on the established linear-quadratic regulator procedures.

As already mentioned, the linear-quadratic methodology does not allow for direct consideration of constraints (2.38). The control constraints, therefore, will be imposed heuristically after calculation of the feedback law. Clearly, this heuristic consideration of control constraints will potentially lead to suboptimal control results. The impact of this (and further) simplification(s) on the control objectives will be ultimately judged by simulation through comparison with the nonlinear optimal control results.

5.2 Linear-Quadratic Formulation of the Sewer Network Control Problem

The linear-quadratic methodology is not directly applicable if there are time delays, like those appearing in the process model (2.37). This difficulty may be readily circumvented by introducing some auxiliary variables $\tilde{x}_i$ (Marinaki and Papageorgiou, 1996a). Thus, for example, for a control variable u_j appearing in the model equations with time delay κ_{uj}, one may introduce the auxiliary additional state equations

$$\tilde{x}_1(k+1) = u_j(k)$$

$$\tilde{x}_2(k+1) = \tilde{x}_1(k) \tag{5.1}$$

$$\ldots$$

$$\tilde{x}_{\kappa_{uj}}(k+1) = \tilde{x}_{\kappa_{uj}-1}(k)$$

and substitute $\tilde{x}_{\kappa_{uj}}(k)$ in all model equations where $u_j(k - \kappa_{uj})$ appears.

This modification can be performed for all time-delayed control and state variables of the process model. The auxiliary variables $\tilde{x}$ are regarded as additional state variables that are incorporated in the state vector $\mathbf{x}$. With this modification, (2.37) obtains the simpler form

$$\mathbf{x}(k+1) = \mathbf{f}[\mathbf{x}(k), \mathbf{u}(k), \mathbf{d}(k)]. \tag{5.2}$$

To facilitate the application of linear controller design, linearization around a stationary nominal point is required. For the definition of this point, a nominal rainfall is considered that leads to constant nominal external inflow values $\mathbf{d}^N$ such that $d_1^N + d_2^N + \ldots + d_{nd}^N = r_{max}$ results in absence of any control actions (all gates opened). Under nominal conditions no overflows occur, because we have assumed that the sum of external inflows equals r_{max}. Consequently, from equation (2.32), we obtain for each reservoir's outflow $u_i^N = u_{in,i}^N$. The nominal reservoir storages V_i^N are provided by inversion of (2.32), taking $u_{un,i}(V_i(k))$ equal to u_i^N, whereas the nominal values of the link outflows q_i^N and of the auxiliary variables $\tilde{x}_i^N$ are readily obtained from d_i^N, u_i^N and x_i^N via stationary continuity considerations. The nominal steady state just described corresponds to a steady-state form of (5.2):

$$\mathbf{x}^N = \mathbf{f}(\mathbf{x}^N, \mathbf{u}^N, \mathbf{d}^N). \tag{5.3}$$

Linearization of (5.2) around this steady state leads to

$$\Delta x(k+1) = \frac{\partial \mathbf{f}}{\partial \mathbf{x}}\Big|_N \Delta x(k) + \frac{\partial \mathbf{f}}{\partial \mathbf{u}}\Big|_N \Delta u(k) + \frac{\partial \mathbf{f}}{\partial \mathbf{d}}\Big|_N \Delta d(k) \qquad (5.4)$$

where $\Delta x(k) = x(k) - x^N$, $\Delta u(k) = u(k) - u^N$, and $\Delta d(k) = d(k) - d^N$ are the linearized variables and $\mathbf{A} = \partial \mathbf{f}/\partial \mathbf{x}\big|_N$, $\mathbf{B} = \partial \mathbf{f}/\partial \mathbf{u}\big|_N$, $\mathbf{C} = \partial \mathbf{f}/\partial \mathbf{d}\big|_N$ are the time-invariant state, control, and disturbance matrices of the linearized system with appropriate dimensions.

If $\Delta d(k)$ vanishes, a feedback law without feedforward terms will be obtained, and this control law will react to the manifest impact of inflows on the measurable storages, rather than to inflow forecasts. On the other hand, if sufficiently accurate inflow forecasts are available, it is possible to incorporate them in the problem formulation ($\Delta d(k) \neq 0$) and, thus, to design a control law with additional feedforward terms to anticipate to some extent the impact of future inflows. An extended design procedure, that explicitly considers inflow forecasts is presented in Sections 5.3.3 and 5.3.4.

The design of a feedback law in case where no inflow predictions are used does not require a controllable linear model. However, the design of a feedback law that enables direct consideration of inflow predictions requires a stationary solution $\mathbf{P}$ of the Ricatti equation (see Section 5.3.4) that is available only for controllable systems (for uncontrollable systems only a stationary gain matrix $\mathbf{L}$ may be obtained). A linear system is controllable (Dorato et al., 1995) if and only if (for nonsingular $\mathbf{A}$),

$$\text{rank}[\mathbf{B}|\mathbf{A}\mathbf{B}|\mathbf{A}^2\mathbf{B}|\ldots|\mathbf{A}^{n-1}\mathbf{B}] = n. \qquad (5.5)$$

where n is the number of state variables.

To obtain a *controllable* linear model, the n_x state variables and according state equations, corresponding to the n_x reservoirs (2.32), are replaced by $n_x - 1$ alternative state variables and state equations. The new state variables and state equations are obtained by building $n_x - 1$ independent differences of the old state equations. For example, if the linearized conservation equations of reservoirs i and j are

$$\Delta V_i(k+1) = \Delta V_i(k) - \Delta u_i(k) + \Delta d_i(k) \qquad (5.6)$$

$$\Delta V_j(k+1) = \Delta V_j(k) - \Delta u_j(k) + \Delta u_l(k) \qquad (5.7)$$

respectively, a new state equation may be obtained with new state variable

$$\Delta x_i'(k+1) = \frac{\Delta V_i(k+1)}{V_{i,max} - V_i^N} - \frac{\Delta V_j(k+1)}{V_{j,max} - V_j^N} \qquad (5.8)$$

if (5.6) and (5.7) are divided by $V_{i,max} - V_i^N$ and $V_{j,max} - V_j^N$, respectively, and subtracted from each other. Note that the modification (5.8) is applied only to the reservoir state equations, whereas the other state equations (for the link outflows and for the auxiliary variables) remain unchanged.

Of course, there are many different ways of building n_x-1 independent differences (5.8) out of n_x original state equations. One way is to take the differences of subsequent reservoirs, that is, build each new $\Delta x_i'$ from the old ΔV_i and ΔV_{i+1}, $i = 1,...,n_x-1$. Another possibility is to consider a fixed reference reservoir j and to build each new $\Delta x_i'$ from the old ΔV_j and ΔV_i for $i = 1,...,n_x$, $i \neq j$. For convenience, Δx will be used in the sequel of the monograph to denote the new state variable $\Delta x'$.

This formulation provides a suitable basis for the inclusion of inflow forecasts in the control law because of the *controllability* of the corresponding linear model, as the controllability condition (5.5) is satisfied for the new model.

A quadratic criterion that considers the control objectives mentioned in Section 3.1 has the general form

$$J = \sum_{k=0}^{\infty} (||\Delta x(k)||_Q^2 + ||\Delta u(k)||_R^2) \tag{5.9}$$

where $||\eta||_S^2 = \eta^T S \eta$ is the quadratic norm of a vector η, and Q and R are nonnegative definite, diagonal weighting matrices. The infinite time horizon in (5.9) is taken in order to obtain a time-invariant feedback law (Section 5.3.4) according to the linear-quadratic optimization theory (Papageorgiou, 1996).

Because of the definition of $\Delta x(k)$, the first term in (5.9) penalizes relative storage differences between reservoirs. The diagonal elements of Q corresponding to the reservoir storages $\Delta x_i'$ are set equal to 1, whereas the diagonal elements of Q corresponding to the link outflows Δq_i and those corresponding to the auxiliary variables $\Delta \tilde{x}_i$ are set equal to zero. A controller designed to minimize this criterion will automatically tend to equalize the relative storage distribution between reservoirs. In other words, the regulator will tend to fill and to empty the reservoirs simultaneously, thereby minimizing the overflowing of some reservoirs while others are not sufficiently filled. This is an indirect way of achieving overflow minimization for the sewer network.

By the choice of the weighting matrix R, that is, its diagonal elements, the magnitude of the control reactions can be influenced. This is necessary in order to avoid high feedback parameters that would lead to nervous control behaviour. Moreover, it provides the possibility to consider, to a certain extent, indirectly, the constraints (2.38), because increased values of the diagonal weighting parameters will lead to lower deviations of outflows from their nominal values. However, as mentioned earlier, the strict consideration of the constraints (2.38) is not guaranteed by the quadratic criterion and must be imposed after the feedback law calculations, that is, the control variables must be truncated according to (2.38). The choice of the diagonal matrix R is performed by a *trial-and-error* procedure so as to achieve a satisfactory control behaviour for a given application network. In the trial-and-error procedure we start with some plausible values for the weighting factors R_i and then look at the control results obtained by simulation. If these results are not satisfactory, for example if the reservoir storages are not sufficiently balanced, or if some control variables u_i violate constraints too often, the corresponding parameters R_i are changed accordingly, and simulation is repeated. It should be noted that during the trial-and-error procedure the weighting matrix **Q** remains unchanged as the control results depend on the *relative* magnitude of the weighting matrices **Q** and **R** and not on their absolute values.

As already mentioned, the inflow r(k) into the treatment plant should be as high as possible in order to empty the network as quickly as possible. Therefore, r(k) is not included in the control vector **u**, but is set $r(k) = r_{max}$ and is truncated afterward if necessary to satisfy (2.36).

5.3 Multivariable Control Law

5.3.1 General Problem Formulation

As detailed in Section 4.1, a direct way of considering the main objectives of a sewer network flow control system is via formulation of a suitable nonlinear cost function that is minimized taking into account the state equation and the constraints. In this chapter an indirect, much simpler, approach is taken that leads to a quadratic criterion to be minimized, taking into account the linearized state equation of Section 5.2. In the following, the general problem formulation and the derivation of a multivariable feedback control with and without feedforward terms is derived using the linear-quadratic optimization theory.

A quadratic cost functional in Δx and Δu reads

$$J = \frac{1}{2}\left\|\Delta x(K)\right\|_S^2 + \frac{1}{2}\sum_{k=0}^{K-1}\left(\left\|\Delta x(k)\right\|_Q^2 + \left\|\Delta u(k)\right\|_R^2\right) \qquad (5.10)$$

where $S \geq 0$, $Q \geq 0$, and $R > 0$ are time-invariant, symmetric weighting matrices, and K is the optimization time horizon defined in Section 4.1. (Matrices **A**, **B**, **C**, **Q**, and **R** may be time varying without any change in the equations of Sections 5.3.1, 5.3.2 and 5.3.3, for a time-invariant solution (Section 5.3.4); however, these matrices must be assumed constant.)

For given disturbance trajectory $\Delta d(k)$, $k = 0,...,K-1$, we are looking for the control $\Delta u(k)$ that minimizes (5.10) subject to (5.4) with initial condition

$$\Delta x(0) = \Delta x_0. \qquad (5.11)$$

5.3.2 Necessary Optimality Conditions

To solve the optimization problem of Section 5.3.1, the discrete-time Hamiltonian function (4.14) is given by

$$H[\Delta x(k), \Delta u(k), \lambda(k+1), k] = \frac{1}{2}[\|\Delta x(k)\|_Q^2 + \|\Delta u(k)\|_R^2] +$$
$$+\lambda(k+1)^T[A\Delta x(k) + B\Delta u(k) + C\Delta d(k)] \quad (5.12)$$

where $\lambda(k+1) \in R^n$ are the *Lagrange* multipliers for the corresponding equality constraints (5.4) (Section 4.3.1).

For the formulated discrete-time optimal control problem, the following necessary conditions of optimality hold for $k = 0,...,K-1$ (notation: $x_y = \partial x/\partial y$):

$$\Delta x(k+1) = H_{\lambda(k+1)} = A\Delta x(k) + B\Delta u(k) + C\Delta d(k) \qquad (5.13)$$

$$\lambda(k) = H_{\Delta x(k)} = Q\Delta x(k) + A^T\lambda(k+1) \qquad (5.14)$$

$$\lambda(K) = S\Delta x(K) \qquad (5.15)$$

$$H_{\Delta u(k)} = 0 = R\Delta u(k) + B^T\lambda(k+1). \qquad (5.16)$$

5.3.3 Time-Variant Solution

To find a solution of the formulated problem we start by assuming that

$$\lambda(k) = P(k)\Delta x(k) + z(k) \qquad (5.17)$$

where $P(k)$ is a symmetric n×n matrix and $z(k)$ is a n×1 vector; replacing (5.17) in (5.16) we have

$$R\Delta u(k) + B^T P(k+1)\Delta x(k+1) + B^T z(k+1) = 0. \qquad (5.18)$$

Replacing $\Delta x(k+1)$ in (5.18) from (5.13) and solving for $\Delta u(k)$ we have

$$\Delta u(k) = -[R + B^T P(k+1)B]^{-1}B^T P(k+1)A\Delta x(k)$$
$$- [R + B^T P(k+1)B]^{-1}B^T[P(k+1)C\Delta d(k) + z(k+1)]. \qquad (5.19)$$

Introducing the m×n matrix

$$D(k) = [R + B^T P(k+1)B]^{-1}B^T \qquad (5.20)$$

the m×n gain matrix

$$L(k) = [\mathbf{R} + \mathbf{B}^T\mathbf{P}(k+1)\mathbf{B}]^{-1}\mathbf{B}^T\mathbf{P}(k+1)\mathbf{A} \tag{5.21}$$

and the n-dimensional vector

$$\mathbf{p}(k+1) = \mathbf{P}(k+1)\mathbf{C}\Delta\mathbf{d}(k) + \mathbf{z}(k+1) \tag{5.22}$$

we obtain for the control $\Delta\mathbf{u}(k)$ from (5.19)

$$\Delta\mathbf{u}(k) = -\mathbf{L}(k)\cdot\Delta\mathbf{x}(k) - \mathbf{D}(k)\mathbf{p}(k+1). \tag{5.23}$$

Replacing (5.13), (5.16) and (5.17) in (5.14) we obtain

$$\mathbf{P}(k)\Delta\mathbf{x}(k) + \mathbf{z}(k) = \mathbf{Q}\Delta\mathbf{x}(k) + \mathbf{A}^T\mathbf{P}(k+1)\,[\mathbf{A}\Delta\mathbf{x}(k) + \mathbf{B}\Delta\mathbf{u}(k)] \\ + \mathbf{A}^T[\mathbf{P}(k+1)\mathbf{C}\Delta\mathbf{d}(k) + \mathbf{z}(k+1)]. \tag{5.24}$$

Using (5.22) and (5.23) in (5.24) we obtain after some manipulations

$$\mathbf{P}(k)\Delta\mathbf{x}(k) + \mathbf{z}(k) = [\mathbf{Q} + \mathbf{A}^T\mathbf{P}(k+1)\mathbf{A} - \mathbf{A}^T\mathbf{P}(k+1)\mathbf{B}\mathbf{L}(k)]\Delta\mathbf{x}(k) \\ + \mathbf{A}^T[\mathbf{I} - \mathbf{P}(k+1)\mathbf{B}\mathbf{D}(k)]\mathbf{p}(k+1). \tag{5.25}$$

By equalizing the coefficients in (5.25) we have

$$\mathbf{P}(k) = \mathbf{Q} + \mathbf{A}^T\mathbf{P}(k+1)\mathbf{A} - \mathbf{A}^T\mathbf{P}(k+1)\mathbf{B}\mathbf{L}(k) \tag{5.26}$$

$$\mathbf{z}(k) = \mathbf{A}^T[\mathbf{I} - \mathbf{P}(k+1)\mathbf{B}\mathbf{D}(k)]\mathbf{p}(k+1). \tag{5.27}$$

Using (5.27) and (5.22) we obtain for the vector $\mathbf{p}(k)$

$$\mathbf{p}(k) = \mathbf{P}(k)\mathbf{C}\Delta\mathbf{d}(k-1) + \mathbf{Z}(k)\mathbf{p}(k+1) \tag{5.28}$$

where

$$\mathbf{Z}(k) = \mathbf{A}^T[\mathbf{I} - \mathbf{P}(k+1)\mathbf{B}\mathbf{D}(k)]. \tag{5.29}$$

Finally, from the terminal condition (5.15) we have with (5.17)

$$\mathbf{S}\Delta\mathbf{x}(K) = \mathbf{P}(K)\Delta\mathbf{x}(K) + \mathbf{z}(K) \tag{5.30}$$

which is satisfied for

$$\mathbf{P}(K) = \mathbf{S} \tag{5.31}$$

and $\mathbf{z}(K) = \mathbf{0}$. The latter implies with (5.22)

$$\mathbf{p}(K) = \mathbf{P}(K)\mathbf{C}\cdot\Delta\mathbf{d}(K-1). \tag{5.32}$$

In summary, the time-varying solution of the linear-quadratic problem of Section 5.3.1 is given by the linear feedback law

$$\mathbf{u}(k) = \mathbf{u}^N - \mathbf{L}(k)\cdot\Delta\mathbf{x}(k) - \mathbf{U}(k) \tag{5.33}$$

where

$$\mathbf{U}(k) = \mathbf{D}(k)\mathbf{p}(k+1). \tag{5.34}$$

The time-variant gain matrix $\mathbf{L}(k)$ and Riccati matrix $\mathbf{P}(k)$ may be obtained by backward integration of the interconnected equations (5.21) and (5.26), starting from the terminal condition (5.31). The vector $\mathbf{p}(k)$ is then calculated by backward integration of (5.28), starting from the terminal condition (5.32) with $\mathbf{D}(k)$ and $\mathbf{Z}(k)$ defined by (5.20) and (5.29), respectively, and with the known disturbance trajectory $\Delta\mathbf{d}(k)$, $k = 0,...,K-1$.

It should be noted that equation (5.33) is an *extended linear-quadratic control law* that takes into account the future disturbances (*i.e.*, the inflow predictions for the sewer network control problem). Thus, the second term on the right-hand side of (5.33) is the feedback portion of the control law, whereas the third term may be regarded as a feedforward term, accounting for future disturbances. Clearly, if, for $k = 0,...,K-1$, $\Delta\mathbf{d}(k) = \mathbf{0}$, then from (5.28) and (5.32) $\mathbf{p}(k+1) = \mathbf{0}$, leading to a *purely feedback control law*

$$\mathbf{u}(k) = \mathbf{u}^N - \mathbf{L}(k)\cdot\Delta\mathbf{x}(k). \tag{5.35}$$

5.3.4 Time-Invariant Solution

For most practical applications, a time-invariant solution with regard to the feedback terms is preferable. To obtain a time-invariant feedback solution, we make the following assumptions:

- The time horizon is infinite: $K\rightarrow\infty$
- The system $[\mathbf{A}, \mathbf{B}]$ is controllable
- The system $[\mathbf{A}, \mathbf{F}]$ is observable, where $\mathbf{F}$ is any matrix such that $\mathbf{F}^T\mathbf{F} = \mathbf{Q}$.

Under these assumptions the backward integration of the Riccati matrix $\mathbf{P}(k)$, starting from any terminal condition $\mathbf{P}(K) \geq \mathbf{0}$, converges toward a unique stationary value $\mathbf{P} \geq \mathbf{0}$. If the controllability assumption is not satisfied, the interconnected equations (5.21) and (5.26) may still converge toward a stationary gain matrix $\mathbf{L}$, but possibly not toward any stationary value for the Riccati matrix $\mathbf{P}$. For the sewer network control problem, these assumptions hold (see Section 5.2) and, thus, by using the terminal condition $\mathbf{P}(K)=\mathbf{I}$, the time-invariant matrices $\mathbf{L}$ and $\mathbf{P}$ may be obtained.

Given $\mathbf{L}$, $\mathbf{P}$, we may calculate the corresponding stationary matrices $\mathbf{D}$, $\mathbf{L}$, $\mathbf{Z}$ via (5.20), (5.21), (5.29), respectively, and the control law is given by

$$\mathbf{u}(k) = \mathbf{u}^N - \mathbf{L}\cdot\Delta\mathbf{x}(k) - \mathbf{U}(k) \qquad (5.36)$$

where the time-variant (feedforward) vector $\mathbf{U}(k)$ is calculated in real time from

$$\mathbf{U}(k) = \mathbf{Dp}(k+1). \qquad (5.37)$$

Clearly, for $K\to\infty$ it is not realistic to assume availability of disturbance trajectories $\Delta\mathbf{d}(k)$ over infinite time. However, if disturbance trajectories are available over the period $k = 0,....,K_s-1$, where K_s is the prediction horizon (in the case of sewer network control, K_s corresponds to the horizon of the real-time available accurate inflow predictions K_p, taken from a predictive rainfall-runoff model, plus the inflow predictions obtained by the use of the prolongation scheme as explained in Section 4.4), then we may set $\Delta\mathbf{d}(k) = \mathbf{0}$, $k = K_s$, $K_s+1,...$. Then, from (5.28) and (5.32) we have $\mathbf{p}(k) = \mathbf{0}$, $k = K_s+1$, $K_s+2,...$, and

$$\mathbf{p}(k) = \mathbf{PC}\,\Delta\mathbf{d}(k-1) + \mathbf{Zp}(k+1) \qquad (5.38)$$

for $k = 1,...,K_s$. Equation (5.38) may be integrated backward (in real time) to provide $\mathbf{p}(k+1)$ and eventually $\mathbf{U}(k)$, $k = 0,...,K_s-1$, that are needed in the control law (5.36).

On the other hand, the purely feedback control law is given by

$$\mathbf{u}(k) = \mathbf{u}^N - \mathbf{L}\cdot\Delta\mathbf{x}(k). \qquad (5.39)$$

It is important to underline that the calculation of $\mathbf{L}$ and $\mathbf{P}$ through backward integration of (5.21) and (5.26) until convergence, may be time consuming for problems with high dimension. However, this computational effort is required only once, off-line.

It must be stressed that the linear-quadratic optimization methodology should be viewed as a vehicle for deriving an efficient gain matrix $\mathbf{L}$, that is, an efficient multivariable feedback regulator, rather than as an attempt to optimize a physically meaningful criterion subject to accurate modelling equations and constraints. Nevertheless, the careful definition of the linear-quadratic problem is expected to lead to good control results, but this should be assessed with simulation investigations before field implementation.

A final comment concerns the real-time measurements required in the feedback regulator. The multivariable regulator (5.36) [or (5.39)] is a so-called *state feedback regulator*, that is, it requires availability of measurements for *all* state variables in real time. In the sewer network context, measurements are typically available for the reservoir storages and possibly for some link outflows, but not necessarily for the retarded auxiliary variables. Thus, if full real-time measurements are lacking, some sort of state estimator (also known as state observer) may have to be developed to estimate the missing measurements in real time. The state estimator is a system driven by both the system input and output that produces as its output estimates of the system states (Anderson and Moore, 1990). As no estimator is normally perfectly accurate, feedback laws using estimates rather than the real measurements of the state will only approximate the ideal situation.

In summary, the application of the control law (5.36) that takes into account real-time measurements and inflow predictions requires the following calculations:

(i) Off-line calculations:
- Calculation of the stationary Riccati matrix $\mathbf{P}$ and gain matrix $\mathbf{L}$
- Calculation of the matrices $\mathbf{D}$ and $\mathbf{Z}$.

(ii) On-line calculations at each time instant k for given predictions $\Delta\mathbf{d}(\kappa)$, $\kappa = k,...,k + K_s - 1$, and given real-time measurements $\Delta\mathbf{x}(k)$:
- Calculation of the time-variant vector $\mathbf{p}(\kappa)$, $\kappa = k + 1,...,k + K_s$.
- Calculation of $\mathbf{U}(k)$.
- Calculation of $\mathbf{u}(k)$.

Thus, the on-line calculation load is clearly increased compared with the purely feedback regulation (5.39), but is still feasible in real time, as it will be seen in the following section.

5.4 Computational Effort

The application of the multivariable feedback regulator with and without feedforward terms requires off-line and on-line calculations as mentioned in Section 5.3.4. For the particular sewer network control problem considered here, where matrices with high dimensions (see Section 6.4) are treated, the computation time needed for the off-line calculations is fairly high, but for the on-line calculations is quite low (few seconds), thus permitting real-time application of the multivariable feedback controller.

More precisely the off-line calculation of the stationary Riccati matrix $\mathbf{P}$ and gain matrix $\mathbf{L}$ in a Workstation HP 700 requires a computation time on the order of 6 hours. The off-line calculation of the matrices $\mathbf{D}$ and $\mathbf{Z}$ is effectuated in a few seconds, as it requires some matrix operations to be performed only once. The on-line calculations of the time variant vectors $\mathbf{p}(k)$ and $\mathbf{U}(k)$, and the control variables $\mathbf{u}(k)$ are effectuated in few seconds, too. It should be noted that for the feedback regulator without feedforward terms, the computation time needed for the on-line calculation of (5.39) is even less, as the time-variant vectors $\mathbf{p}(k)$ and $\mathbf{U}(k)$ are not calculated and the control law is simpler.

Chapter 6
Application Example

6.1 Application Network

To assess the efficiency of the methodologies described in Chapters 4 and 5 in reducing the overflows and more generally in satisfying the control objectives described in Section 3.1 when applied to a real sewer network, an extended investigation was performed for the sewer network of Obere Iller (in Bavaria, Germany). This network connects five neighboring cities to one treatment plant. The simplified model of this network is depicted in Figure 6.1. In this network, reservoir 7 is a storage element created by setting up a control gate for regulating the flow at the end of a voluminous sewer in the network without overflow capability. There is, however, for emergency needs a bypass of the control gate (a weir over the gate), so that in case of an overload, an overflow $q_{over,7}$ is created that enters the sewer 5 through nodes 4 and 5 (Figure 6.1).

The continuity equations of the simplified model (2.32) that correspond to the eleven reservoirs of the network, are given by:

$$x_1(k+1) = x_1(k) + T(d_1(k) - u_1(k) - q_{over,1}(k)) \tag{6.1}$$

$$x_2(k+1) = x_2(k) + T(d_3(k) - u_2(k) - q_{over,2}(k)) \tag{6.2}$$

$$x_3(k+1) = x_3(k) + T(d_5(k) - u_3(k) - q_{over,3}(k)) \tag{6.3}$$

$$x_4(k+1) = x_4(k) + T(d_6(k) - u_4(k) - q_{over,4}(k)) \tag{6.4}$$

$$x_5(k+1) = x_5(k) + T(d_7(k) - u_5(k) - q_{over,5}(k)) \tag{6.5}$$

$$x_6(k+1) = x_6(k) + T(d_{13}(k) + x_{14}(k) - u_6(k) - q_{over,6}(k)) \tag{6.6}$$

$$x_7(k+1) = x_7(k) + T(x_{15}(k) - u_7(k) - q_{over,7}(k)) \tag{6.7}$$

$$x_8(k+1) = x_8(k) + T(d_8(k) - u_8(k) - q_{over,8}(k)) \tag{6.8}$$

$$x_9(k+1) = x_9(k) + T(d_{10}(k) - u_9(k) - q_{over,9}(k)) \tag{6.9}$$

$$x_{10}(k+1) = x_{10}(k) + T(d_{11}(k) - u_{10}(k) - q_{over,10}(k)) \tag{6.10}$$

$$x_{11}(k+1) = x_{11}(k) + T(x_{17}(k) - u_{11}(k) - q_{over,11}(k)). \tag{6.11}$$

For the sewer outflows $q_1,...,q_6$ Equation (2.10) is used for the simplified model where:

$$q_{u,1}(k) = d_2(k - \kappa_1) + u_1(k - \kappa_1) \tag{6.12}$$

$$q_{u,2}(k) = u_2(k - \kappa_2) + d_4(k - \kappa_7) + q_1(k - \kappa_2) \tag{6.13}$$

$$q_{u,3}(k) = u_5(k - \kappa_3) \tag{6.14}$$

$$q_{u,4}(k) = u_6(k - \kappa_4) + u_3(k - \kappa_8) + u_4(k - \kappa_8) + q_2(k - \kappa_8) \tag{6.15}$$

$$q_{u,5}(k) = u_7(k - \kappa_5) + u_8(k - \kappa_9) + d_9(k - \kappa_5) + u_9(k - \kappa_9)$$

$$+ q_{over,7}(k - \kappa_5) \tag{6.16}$$

$$q_{u,6}(k) = d_{12}(k - \kappa_6) + u_{10}(k - \kappa_6) + q_5(k - \kappa_6). \tag{6.17}$$

The constraints of the reservoir outflows are given by (2.34), whereas from (2.12), we have for this network the specific constraints:

$$u_1(k) \le q_{max,1} - d_2(k) \tag{6.18}$$

$$u_2(k) \le q_{max,2} - q_1(k) - d_4(k + \kappa_2 - \kappa_7) \tag{6.19}$$

$$u_5(k) \le q_{max,3} \tag{6.20}$$

$$u_3(k - \kappa_8 + \kappa_4) + u_4(k - \kappa_8 + \kappa_4) + u_6(k) \le q_{max,4} - q_2(k - \kappa_8 + \kappa_4) \tag{6.21}$$

$$u_7(k - \kappa_5 + \kappa_9) + u_8(k) + u_9(k) \le q_{max,5} - d_9(k - \kappa_5 + \kappa_9)$$

$$- q_{over,7}(k - \kappa_5 + \kappa_9) \tag{6.22}$$

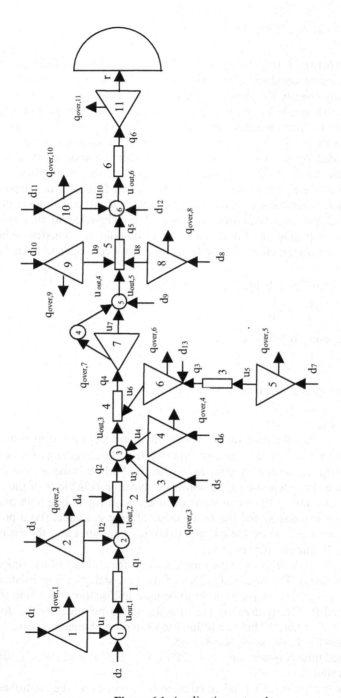

Figure 6.1. Application network.

$$u_{10}(k) \le q_{max,6} - q_5(k) - d_{12}(k). \tag{6.23}$$

Constraints (6.18), (6.19), (6.20), and (6.23) are regarded as additional upper bounds for the control variables $u_1(k)$, $u_2(k)$, $u_5(k)$, and $u_{10}(k)$, respectively.

To empirically specify the values of the time delays and the time constants in the sewer links, the simulation program KANSIM was used. Positive and negative step responses of the flow process were calculated for the sewers 1,...,6 by using step inputs at various levels of flow and recording the corresponding outflow responses provided by KANSIM. The final values for the time delay and time constant of each sewer stretch was then taken equal to the mean value of the corresponding parameter values over all responses. Three levels for the step inputs were used, namely when the sewers are almost empty, half full and almost full.

In Figure 6.2 the step responses for sewer 1 are depicted to illustrate the procedure. By imposing a linear first-order time-delay transfer function (Papageorgiou and Messmer, 1989) we have for each step response from Figure 6.2 (in minutes):

- For step input from $0.08\,\mathrm{m^3/s}$ to $0.3\,\mathrm{m^3/s}$:
  $\kappa_1^+ = 42,\ \tau_1^+ = 14$
  $\kappa_1^- = 34,\ \tau_1^- = 30$
- For step input from $0.3\,\mathrm{m^3/s}$ to $0.6\,\mathrm{m^3/s}$:
  $\kappa_1^+ = 31,\ \tau_1^+ = 15$
  $\kappa_1^- = 27,\ \tau_1^- = 22$
- For step input from $0.6\,\mathrm{m^3/s}$ to $0.9\,\mathrm{m^3/s}$:
  $\kappa_1^+ = 26,\ \tau_1^+ = 18$
  $\kappa_1^- = 25,\ \tau_1^- = 20$

where κ_1^+ and κ_1^- are the time delays for positive and negative step responses, respectively, and τ_1^+ and τ_1^- are the first-order system time constants for positive and negative step responses, respectively. Clearly both parameter values are different for each time step due to the nonlinear dynamic behaviour of the sewer flow process. Thus, taking the mean value of time delays (including both positive and negative step responses) and the mean value of time constants (both positive and negative responses) the estimated time delay is 31 minutes and the estimated time constant is 20 minutes for sewer 1.

In this way, the time delays and the time constants for all links of the simplified model were estimated. The obtained values of the time delays are translated into according numbers of time steps, in order to be used in Equations (6.12) to (6.17), (6.19), (6.21) and (6.22), by dividing the time delays (in minutes) by the discrete time interval T = 3 minutes. Thus the following values are obtained: $\kappa_1=10$, $\kappa_2=16$, $\kappa_3=28$, $\kappa_4=5$, $\kappa_5=10$, $\kappa_6=9$, $\kappa_7=9$, $\kappa_8=8$, $\kappa_9=8$.

The estimated time constants are: $\tau_1 = 1200$ s, $\tau_2 = 2220$ s, $\tau_3=2100$ s, $\tau_4=1920$s, $\tau_5=1800$ s, $\tau_6=1800$s .

The discrete time interval T is taken equal to 180 seconds for the control and 60 seconds for the simulation. For the linear-quadratic formulation of the present problem, the auxiliary variables (5.1) are used to take into account the time delays

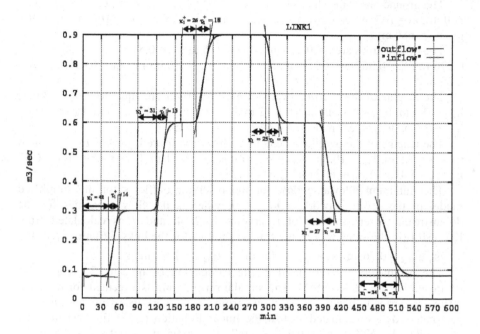

Figure 6.2. Step responses for sewer 1.

as mentioned in Section 5.2, and the corresponding equations are attached to the original state equations. Thus, taking into account the time delays, we have 159 state variables for the present problem (10 for the reservoirs, 6 for the link outflows, and 143 for the auxiliary variables).

The flow capacities $q_{max,i}$ of the network links may be estimated by use of the KANSIM program to correspond to the sewer flows when water level equals the sewer height. This, however, may be a conservative estimation for sewers with flow under pressure (*e.g.*, due to a full upstream reservoir), in which case the optimal results obtained from the simplified model may be unjustifiably constrained. Note that initially our values of $q_{max,i}$ were indeed chosen conservatively, leading in some cases to reduced performance of the optimal control approach (Marinaki, 2002). Note, also, that the multivariable regulators do not face this problem as they do not consider explicitly any constraints – flow constraints there are imposed by the real world (or the simulation program). The (nonconservative) values of $q_{max,i}$ that were finally used in this specific application correspond to the sewer flows when water level is smaller or equal to 1.5 times the sewer height; these values are $q_{max,1} = 1.1$ m^3/s, $q_{max,2} = 1.7$ m^3/s, $q_{max,3} = 0.45$ m^3/s, $q_{max,4} = 2.7$ m^3/s, $q_{max,5} = 4.5$ m^3/s, $q_{max,6} = 5.45$ m^3/s.

The maximum reservoir storages are $V_{1,max} = 1200$ m^3, $V_{2,max} = 400$ m^3, $V_{3,max} = 685$ m^3, $V_{4,max} = 1485$ m^3, $V_{5,max} = 400$ m^3, $V_{6,max} = 1500$ m^3, $V_{7,max} = 2200$ m^3, $V_{8,max} = 600$ m^3, $V_{9,max} = 500$ m^3, $V_{10,max} = 1600$ m^3, $V_{11,max} = 6800$ m^3.

The ground area of each reservoir is $e_1 = 360$ m^2, $e_2 = 167$ m^2, $e_3 = 167$ m^2, $e_4 = 360$ m^2, $e_5 = 167$ m^2, $e_6 = 361$ m^2, $e_7 = 1158$ m^2, $e_8 = 167$ m^2, $e_9 = 167$ m^2, $e_{10} = 361$ m^2, $e_{11} = 680$ m^2.

The orifice areas are $f_1 = 0.235$ m^2, $f_2 = 0.276$ m^2, $f_3 = 0.81$ m^2, $f_4 = 0.36$ m^2, $f_5 = 0.276$ m^2, $f_6 = 0.367$ m^2, $f_7 = 1$ m^2, $f_8 = 0.622$ m^2, $f_9 = 0.43$ m^2, $f_{10} = 0.418$ m^2, $f_{11} = 2.25$ m^2.

The heights of the overflow weirs of each reservoir are equal to $h_{w,1} = 3.33$ m, $h_{w,2} = 2.4$ m, $h_{w,3} = 4.11$ m, $h_{w,4} = 4.12$ m, $h_{w,5} = 2.4$ m, $h_{w,6} = 4.16$ m, $h_{w,7} = 1.9$ m, $h_{w,8} = 3.6$ m, $h_{w,9} = 2.99$ m, $h_{w,10} = 4.44$ m, $h_{w,11} = 10$ m.

The lengths of the overflow weirs of each reservoir are $l_{w,1} = 8.48$ m, $l_{w,2} = 4.38$ m, $l_{w,3} = 7.5$ m, $l_{w,4} = 10.5$ m, $l_{w,5} = 4.38$ m, $l_{w,6} = 10.61$ m, $l_{w,7} = 3.47$ m, $l_{w,8} = 6.57$ m, $l_{w,9} = 5.47$ m, $l_{w,10} = 11.31$ m, $l_{w,11} = 14.25$ m.

The minimum allowable flow for the reservoir outflows for the simplified model of the sewer network is $u_{min} = 0$, whereas the flow capacities for the downstream sewer stretches of reservoirs 1, 2, 4, and 7 as calculated from KANSIM (in the same way as the flow capacities for the links 1-6) are $u_{cap,1} = 0.8$ m^3/s, $u_{cap,2} = 0.76$ m^3/s, $u_{cap,4} = 1.03$ m^3/s, $u_{cap,7} = 1.62$ m^3/s.

The parameters that are used in the calculation of q_{over} in (2.33) are also estimated by use of KANSIM. The simulation program is executed for different inflow scenarios to obtain the overflows from the network reservoirs and then the parameters d_0 are estimated so that the linear Equation (2.33) fits best the data obtained by KANSIM: $d_{0,1} = 0.9$, $d_{0,2} = 0.7$, $d_{0,3} = 0.95$, $d_{0,4} = 0.9$, $d_{0,5} = 0.6$, $d_{0,6} = 0.95$, $d_{0,7} = 0.1$, $d_{0,8} = 0.95$, $d_{0,9} = 0.95$, $d_{0,10} = 0.95$, $d_{0,11} = 1$.

The parameters used in the calculation of u_{un} via (2.36) are estimated by use of the simulation program in the same way as the parameters d_0, that is, suitable parameters c_0 are selected so that the nonlinear Equation (2.36) fits best the data obtained from KANSIM. It should be noted that these values should not be selected in a conservative way for similar reasons as explained for q_{max} above. The parameter values selected are $c_{0,1} = 2.7$ m$^{0.5}$/s, $c_{0,2} = 2.5$ m$^{0.5}$/s, $c_{0,3} = 2.4$ m$^{0.5}$/s, $c_{0,4} = 3$ m$^{0.5}$/s, $c_{0,5} = 2.4$ m$^{0.5}$/s, $c_{0,6} = 2.4$ m$^{0.5}$/s, $c_{0,7} = 1.35$ m$^{0.5}$/s, $c_{0,8} = 2.0$ m$^{0.5}$/s, $c_{0,9} = 2.4$ m$^{0.5}$/s, $c_{0,10} = 2.5$ m$^{0.5}$/s, $c_{0,11} = 1.6$ m$^{0.5}$/s.

The treatment plant has flow capacity $r_{max} = 2$ m^3/s. The nominal values for the external inflows are calculated so that their sum be equal to r_{max} and in accordance with the catchment area and the population of each region to which the external inflow corresponds: $d_1^N = 0.174$ m^3/s, $d_2^N = 0.074$ m^3/s, $d_3^N = 0.132$ m^3/s, $d_4^N = 0.107$ m^3/s, $d_5^N = 0.201$ m^3/s, $d_6^N = 0.201$ m^3/s, $d_7^N = 0.116$ m^3/s, $d_8^N = 0.066$ m^3/s, $d_9^N = 0.066$ m^3/s, $d_{10}^N = 0.182$ m^3/s, $d_{11}^N = 0.364$ m^3/s, $d_{12}^N = 0.049$ m^3/s, $d_{13}^N = 0.268$ m^3/s.

The nominal values of the reservoir outflows, of the reservoir storages and of the link outflows are calculated from the simulation program KANSIM assuming as external inflows the nominal values of the external inflows and keeping all gates open. Thus, the obtained nominal values for the reservoir outflows are $u_1^N = 0.174$ m^3/s, $u_2^N = 0.132$ m^3/s, $u_3^N = 0.201$ m^3/s, $u_4^N = 0.201$ m^3/s, $u_5^N = 0.116$ m^3/s,

$u_6^N = 0.384$ m^3/s, $u_7^N = 1.273$ m^3/s, $u_8^N = 0.066$ m^3/s, $u_9^N = 0.182$ m^3/s, $u_{10}^N =$ 0.364 m^3/s, $u_{11}^N = 2$ m^3/s.

The obtained nominal values for the reservoir storages are $V_1^N = 121$ m^3, $V_2^N =$ 79 m^3, $V_3^N = 19$ m^3, $V_4^N = 11$ m^3, $V_5^N = 48$ m^3, $V_6^N = 325$ m^3, $V_7^N = 833$ m^3, $V_8^N =$ 144 m^3, $V_9^N = 148$ m^3, $V_{10}^N = 377$ m^3, $V_{11}^N = 166$ m^3.

The obtained nominal values for the link outflows are $q_1^N = 0.248$ m^3/s, $q_2^N =$ 0.487 m^3/s, $q_3^N = 0.116$ m^3/s, $q_4^N = 1.273$ m^3/s, $q_5^N = 1.587$ m^3/s, $q_6^N = 2$ m^3/s.

The nominal values for the auxiliary variables are equal to the nominal values of the corresponding variables that have time delays. Thus, for example, for the control variable u_1, that enters (6.12) with time delay κ_1, κ_1 auxiliary variables are used, each of the them having nominal value equal to the nominal value of u_1.

6.2 External Inflows

Three scenarios of external inflows (Figures 6.3, 6.4, and 6.5) are used to investigate the efficacy of the multivariable control law and the nonlinear optimal control for the particular network under different circumstances. The first scenario has locally inhomogeneous inflows, whereas in the second scenario we notice a redistribution with regard to the charge of individual reservoirs. Finally the third scenario has locally and temporally inhomogeneous inflows. These scenarios were created taking into account the extent of the catchment area and the population of the region corresponding to each external inflow, as well as existing real data for the external inflows of reservoirs 3, 4, and 6 (rainfall data recorded in this region between 1961 and 1980).

For the feedback controller without feedforward terms (5.10) no inflow forecasts are required. For the feedback controller with feedforward terms (5.13) as well as for the nonlinear optimal control, availability of accurate predictions for the external inflows $d_1,...,d_{13}$ will be assumed initially in order to investigate their behaviour under ideal conditions. Afterward, the behaviour of both methods will be investigated in case accurate inflow predictions are available only for a part of the time horizon K. Of course, in a real-time environment, future inflows cannot be exactly known, even if a predictive rainfallrunoff model is provided.

6.3 Nonlinear Optimal Control

For the real-time nonlinear optimal control, the method of rolling horizon is used, as was described in Section 4.4, with updated inflow predictions and updated initial conditions. The cost criterion that will be minimized, according to Section 4.1, has the following form:

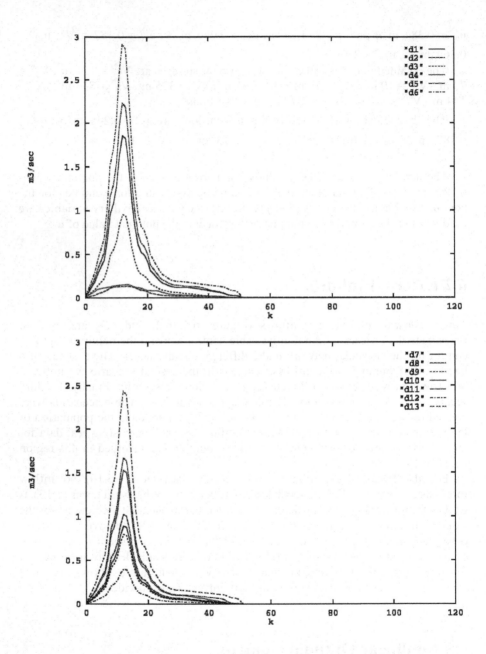

Figure 6.3. External inflows: Scenario 1; T = 180 seconds.

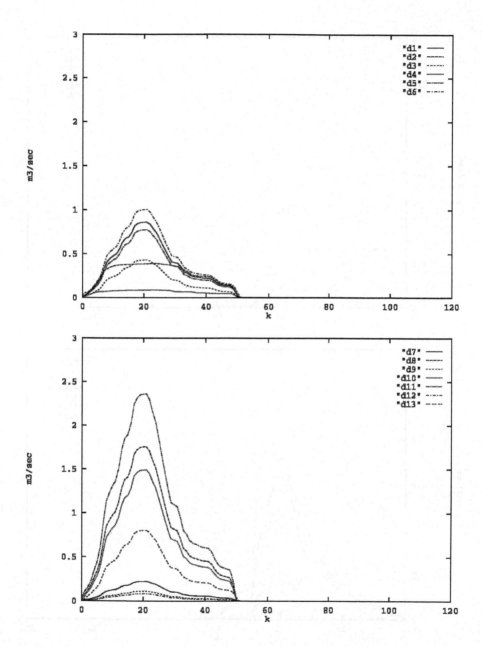

Figure 6.4. External inflows: Scenario 2; T = 180 seconds.

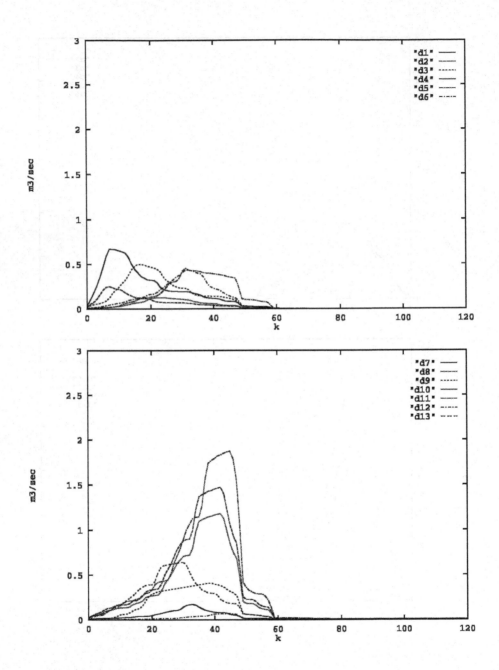

Figure 6.5. External inflows: Scenario 3; T = 180 seconds.

$$J=w_1 \; \psi[V_{7,max} - V_7(K)]^2 + w_2 \sum_{i=1, i\neq 7}^{11} S_i \psi[d_{0,i}(V_{i,max} - V_i(K))/T]^2$$

$$+ w_4 \sum_{i=1}^{11} [V_i(K) - v_i V_G(K)]^2$$

$$+ \sum_{k=0}^{K-1} \left\{ w_1 \psi[V_{7,max} - V_7(k)]^2 + w_2 \sum_{i=1, i\neq 7}^{11} S_i \psi[d_{0,i}(V_{i,max} - V_i(k))/T]^2 + \right.$$

$$+ w_3 [r_{max} - r(k)]^2 + w_4 \sum_{i=1}^{11} [V_i(k) - v_i V_G(k)]^2 +$$

$$w_5 \sum_{i=1}^{11} Q_i [u_i(k) - u_i(k-1)]^2 +$$

$$+ w_6 \psi[q_{max,4} - u_3(k-\kappa_8+\kappa_4) - q_2(k-\kappa_8+\kappa_4) - u_4(k-\kappa_8+\kappa_4) - u_6(k)]^2 +$$

$$+ w_7 \psi[q_{max,5} - u_7(k-\kappa_5+\kappa_9) - u_8(k) - u_9(k) - d_9(k-\kappa_5+\kappa_9)$$

$$\left. - q_{over,7}(k-\kappa_5+\kappa_9)]^2 \right\} \tag{6.24}$$

where the last two penalty terms are used to take into account the constraints (6.21), and (6.22), respectively, and they are not needed if these constraints are taken into account directly (see Section 4.3.7). The weight coefficients S_i and Q_i are set equal to 1.0. The other weight coefficients are

$$w_1 = 0.1, \; w_2 = 1, \; w_3 = 0.5, \; w_4 = 10^{-8}, \; w_5 = 0.1, \; w_6 = 10, \; w_7 = 10.$$

These weight values were chosen after a trial-and-error procedure, taking into account the physical dimensions of the quantities involved and the desired priority established in Section 4.1. The optimization horizon was taken equal to K = 120 (= 6 h) when the nonlinear optimal control was applied to the particular sewer network in order to estimate the weight coefficients to be used in (6.24).

When the rolling horizon concept is applied, the initial reservoir storages, the link outflows, and all retarded variables with negative time arguments are required to solve the mathematical optimization problem in real time. The reservoir storages and the link and reservoir outflows are measurable variables, and all other variables are estimated using the measured variables. For example, the estimation of the κ_2 retarded values of the outflow u_2 to be used in (6.13), if the repetition period is $k_R=1$ and the discrete time intervals of simulation and optimization are equal to 60 seconds and 180 seconds, respectively, is performed as follows:

At time step k=0 the optimization is executed using, for the κ_2 retarded values of u_2, the measured values of u_2 for the time k=0, as these values correspond to a stationary condition. The next time the optimization is repeated, at time step k=3, the mean value of the outflows u_2 obtained at the time steps k=1, k=2, and k=3 is calculated and replaces the value of $u_2(-1)$, whereas the other values, that is, $u_2(-2)$ until $u_2(-\kappa_2)$, remain unchanged, and so forth.

6.4 Linear-Quadratic Formulation

Two multivariable controllers, one with and another without feedforward terms, were designed via the linear-quadratic methodology to investigate both the reactive and anticipatory regulator behaviour for the particular network. The controllers were programmed and were connected as an additional module to the simulation program KANSIM. For the design procedure, reservoir 7 of the particular application network is considered as the reference reservoir j as this reservoir is geographically in the centre of this sewer network. Thus we have as state variables $\Delta x_i = (V_i - V_i^N)/(V_{i,\max} - V_i^N) - (V_7 - V_7^N)/(V_{7,\max} - V_7^N)$, i=1,...,11, i≠7.

Due to the existing long time delays in the particular network considered here, the total number of state variables (for T=180 seconds) is 159, whereas the total number of control and disturbance variables is 10 and 13, respectively. Thus, the dimensions of the matrices **L**, and **P** are 10 × 159 and 159 × 159, respectively, and the vectors **U**, **p** have dimensions 10 and 159, respectively. The involved on-line and off-line computational effort was discussed in Chapter 5.

For a given application network, to achieve satisfactory control results, an appropriate weighting matrix **R** should be selected. By the choice of the weighting matrix **R**, composed of nonnegative diagonal elements r_i, the magnitude of the control reactions can be influenced. This choice is made via a *trial-and-error* procedure (Section 5.2). The weighting matrix **Q** remains unchanged and is selected in our problem as described in Section 5.2.

The analysis of the regulator's behaviour for different weighting matrices **R** illustrates how this matrix affects the control results. For the particular sewer network control problem this analysis is performed for the three scenarios of external inflows presented in Section 6.2. The results show that for smaller values of the diagonal elements r_i (*e.g.*, $r_i=10^{-6}$, i=1,...,10) the second term of the quadratic criterion (5.9) becomes less important, leading to greater control values than in case of higher values of the diagonal elements r_i (*e.g.*, $r_i=10^{-2}$, i=1,...,10), where lower deviations of the control variables from their nominal values are observed. By the same token, the balancing of reservoir storages is stronger for lower values of r_i than for higher values. However, for small values of r_i, instability problems are observed in reservoir 7 due to model mismatch between the design model and the KANSIM simulator. (Note that the design model includes constant time delays for the sewer links whereas the corresponding KANSIM time delays are state-dependent as explained in Section 2.3.1.) These problems are eliminated when the diagonal element of matrix $\mathbf{R}$ corresponding to the outflow of reservoir 7 is given a value greater or equal to 0.1. A weighting matrix $\mathbf{R}$ consisting of diagonal elements $r_i = 10^{-2}$ for $i = 1,\ldots,10$, $i \neq 7$, and $r_7 = 0.1$ is finally selected to be used for the multivariable regulator with and without feedforward terms. This choice leads to very satisfactory relative reservoir storage equalization and to the avoidance of any instability problems.

After the calculation of the control variables from Equations (5.36) or (5.39) for the multivariable regulators with or without feedforward terms, respectively, a water level control scheme (Papageorgiou and Messmer, 1985) is used for reservoir 7 in the control module connected to KANSIM. The control task is to keep the water level in reservoir 7 near the value $h_{w,7}$ by appropriate operation of the control gates of the reservoirs upstream of reservoir 7 and by this way to avoid the overloading of this storage element. To this end, a quantity $\tilde{u}(k)$ is calculated and is used as an upper bound for the sum of outflows of reservoirs 1, 2, 3, 4, and 6. If the sum of these outflows is greater than $\tilde{u}(k)$, a value c, where

$$c = \tilde{u}(k) / \sum_{i=1, i \neq 5}^{6} u_i(k) \tag{6.25}$$

is calculated, and each outflow $u_1(k)$, $u_2(k)$, $u_3(k)$, $u_4(k)$, and $u_6(k)$ is multiplied by c, so that the constraint be respected. The following water level regulator is used for the calculation of $\tilde{u}(k)$:

$$\tilde{u}(k) = K_R \cdot (0.5/T_D) \cdot h_{w,7} \cdot (h_{w,7} - h_7(k)) + p_7(k) - d_2(k) - d_4(k) \tag{6.26}$$

where:
- K_R is equal to $\alpha \pi l / 4$, where α is a regulation coefficient that was chosen equal to 0.615 and l is the length of the sewer representing reservoir 7 which is equal to 1261.5 m.
- $h_7(k)$ is the water level in reservoir 7 (in m).
- $p_7(k)$ (in m^3/s) is the maximum between $u_7(k)$ and $u_{cap,7}$.
- T_D is a parameter experimentally specified (in seconds). More specifically, T_D was determined by taking constant values for $d_2(k)$, $d_4(k)$, and $p_7(k)$ and checking the value of T_D, for which h_7 converges to $h_{w,7}$ faster. For example, taking the constant values $d_2(k) = 0.0132\, m^3/s$, $d_4(k) = 0.192\, m^3/s$, $p_7(k) = u_7(k) = 0.22\, m^3/s$, the simulation horizon equal to 8 hours, the simulation time step equal to 60 seconds, and the control time step equal to 180 seconds, the results depicted in Figure 6.6 are obtained for the particular sewer network. From Figure 6.6 it can be seen that for $T_D=1000$ seconds, $h_7(k)$ converges quite fast to $h_{w,7}$ without significant

overshooting. This value, however, was decreased to T_D = 235 seconds during the general simulation investigations of Chapter 7 as it was found more suitable to protect reservoir 7 from overload.

6.5 Simulation

The simulation program KANSIM, that is based on the accurate model of the sewer network as described in Section 2.3, is used as a basis for testing and comparing the control performance of the multivariable feedback control versus the nonlinear optimal control. This program is also used to simulate the no-control case, so as to assess the achievable improvements via application of efficient central control strategies to the particular network.

The specific simulation program simulates the underlying actions of local direct control as well. More precisely, in this program, the flow from each control gate is kept, if physically feasible, close to specific reference values, that is, the ones provided by the central control. It is assumed that this is done perfectly, that is, that the outflow of each reservoir is equal to the corresponding predefined reference value for the corresponding time interval, as long as there is enough water level difference to produce a flow greater or equal to the reference value. If the water level difference is lower than that limit, that is, if the reservoir is almost empty, the control gate is assumed completely opened and the flow is calculated accordingly (Section 2.4). In the no-control case, the gates are assumed opened to 28%, 27%,

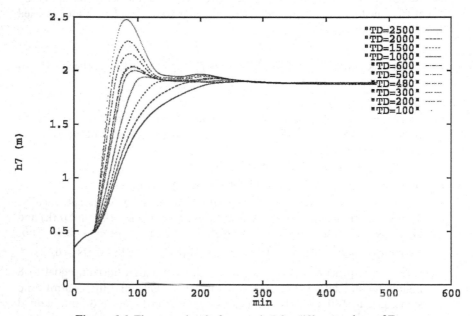

Figure 6.6. The water level of reservoir 7 for different values of T_D.

100%, 50%, and 100% of their maximum opening height for reservoirs 1, 2 to 6, 7, 8 to 10, and 11, respectively, and the flow is calculated accordingly. The selection of the above percentages for the opening heights of the orifices, which have different geometric characteristics (Section 6.1), was performed by conducting many simulation runs using different percentages so as to achieve acceptable fixed-control performance without overloading reservoir 7.

Chapter 7
Simulation Results

7.1 No-Control Case

The calculation of the outflows, the overflows, and the storages of the reservoirs for the sewer network of Obere Iller, when no control actions are taken, is performed using the inflow scenarios presented in Section 6.2 and the simulation program KANSIM. The results obtained constitute the basis for testing and comparing the control strategies applied in this network. The simulation results of the no-control case are presented in Table 7.1 and in Figures 7.1 to 7.9 and are briefly commented on the following discussion.

Figures 7.1, 7.4, and 7.7 display the reservoir outflows for the three investigated scenarios. It should be noted that, although the gate opening of each reservoir is constant in time and equal for all scenarios (Section 6.5), the resulting outflows are time-varying because they depend on the water level in the corresponding reservoirs according to (2.15). Note also that negative flows in the figures correspond to backflow, which may appear if the downstream pressure of a reservoir or sewer stretch is higher than the upstream pressure.

For *scenario 1*, external inflows are stronger upstream of reservoir 7 than downstream of reservoir 7 (Figure 6.3). Reservoirs 4, 6, 3, and 1 receive very strong external inflows (d_6, d_{13}, d_5, d_1, respectively), whereas reservoirs 10 and 9 receive quite strong inflows (d_{11}, d_{10}, respectively); hence, an overflow appears probable for these reservoirs. Indeed, as shown in Figure 7.3 and in Table 7.1, reservoirs 1 to 6 and reservoir 9 overflow, whereas a small overload is created in reservoir 7 in the no-control case.

For *scenario 2*, external inflows are stronger downstream of reservoir 7 than upstream of reservoir 7 (Figure 6.4). Reservoirs 10, 8, and 9 receive very strong external inflows (d_{11}, d_8, d_{10}, respectively), and thus large overflows appear in these reservoirs (Figure 7.6, Table 7.1) when no control actions are taken. Reservoirs 1 and 2 are also overflowing, although they do not have particularly strong external inflows. This is due to the opening height of the gates, which is 28% and 27% of the maximum opening height for reservoirs 1 and 2, respectively. However, it should be noted that the selection of the percentages of the opening heights of the orifices (see Section 6.5) leads to the avoidance of overloading of reservoir 7 for this particular scenario.

For *scenario 3*, which has locally and temporally inhomogeneous inflows, reservoirs 10, 8, and 9 receive very strong external inflows (d_{11}, d_8, d_{10}, respectively), whereby the inflow peaks are at the time periods [114 minutes, 138 minutes], [106 minutes, 130 minutes] and   [106 minutes, 130 minutes], respectively. Thus, these reservoirs are largely overflowing (Table 7.1, Figure 7.9) at the time period of the respective inflow peaks. Reservoir 2 has a small overflow when its inflow (d_3) has a peak due to the relatively low percentage of the opening height of the gate.

It should be noted that for all three scenarios of external inflows, the total available storage volume of the sewer network is not fully utilized in the no-control case. From Figures 7.2, 7.5, and 7.8, it can be seen that there are reservoirs that are not totally filled while others are overflowing.

Table 7.1. Reservoir overflows and overload of reservoir 7 in $[m^3]$ for the no-control case.

Reservoir	Scenario 1	Scenario 2	Scenario 3
1	932	541	0
2	394	72	164
3	353	0	0
4	684	0	0
5	338	0	0
6	647	0	0
8	0	471	220
9	284	1150	697
10	0	2047	966
11	0	0	0
Total	3632	4281	2047
7	96	0	0

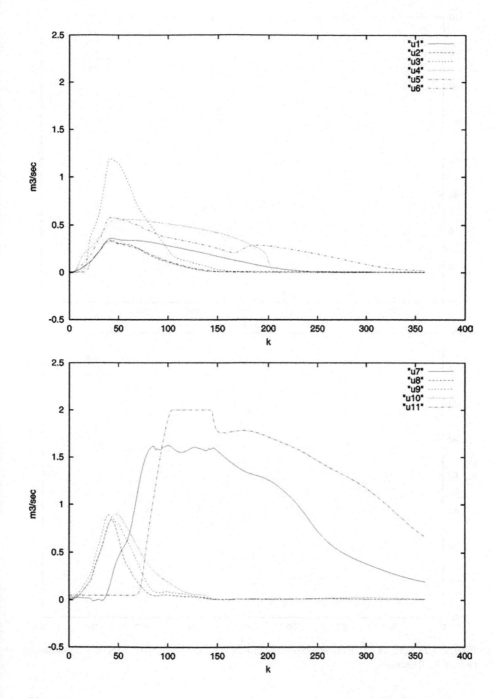

Figure 7.1. Scenario 1: Reservoir outflows $u_i(k)$ for the no-control case; $T = 60$ seconds.

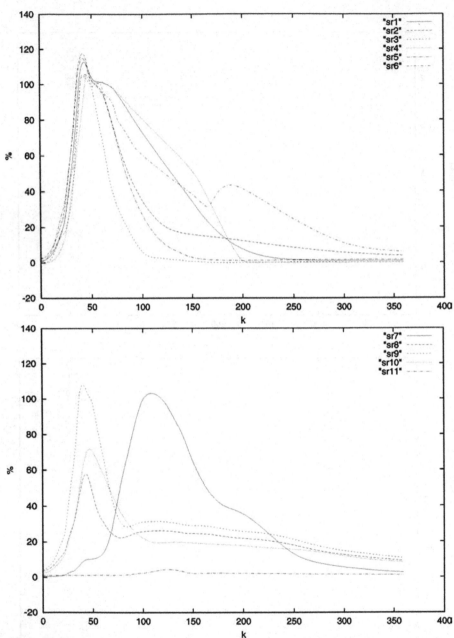

Figure 7.2. Scenario 1: Relative reservoir storages $(V_i(k)/V_{i,max})100\%$ for the no-control case; $T = 60$ seconds.

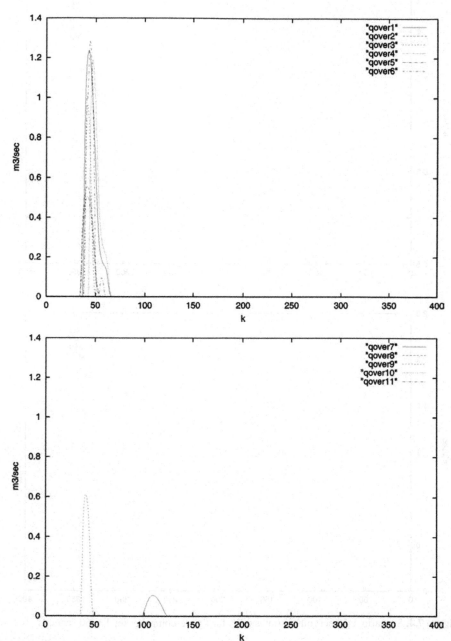

Figure 7.3. Scenario 1: Reservoir overflows $q_{over,i}(k)$ for the no-control case; $T = 60$ seconds.

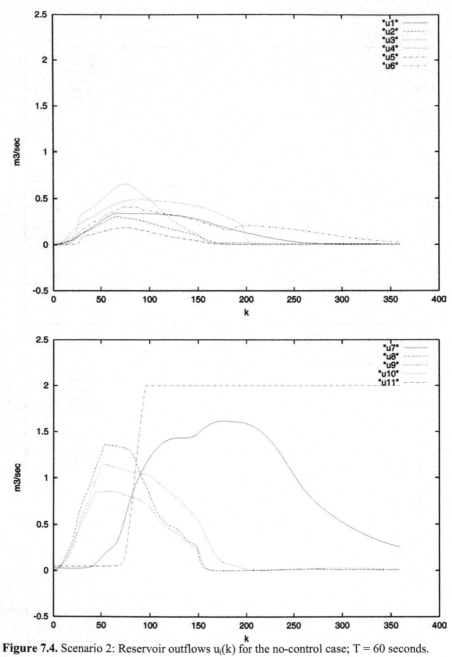

Figure 7.4. Scenario 2: Reservoir outflows $u_i(k)$ for the no-control case; T = 60 seconds.

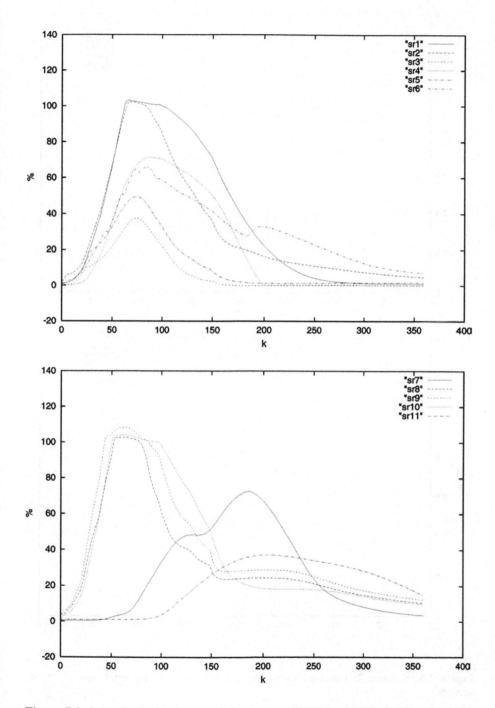

Figure 7.5. Scenario 2: Relative reservoir storages $(V_i(k)/V_{i,max})100\%$ for the no-control case; $T = 60$ seconds.

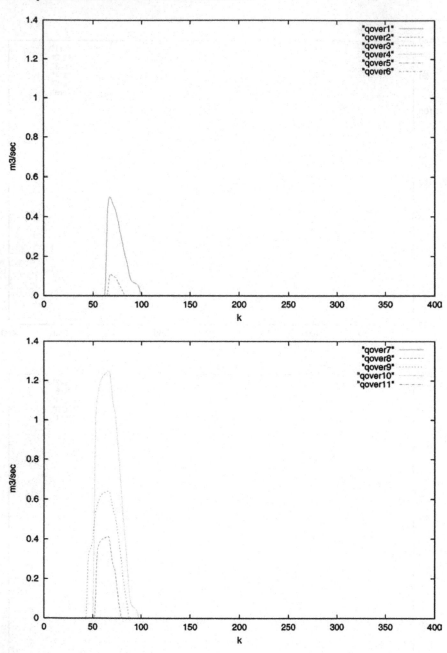

Figure 7.6. Scenario 2: Reservoir overflows $q_{over,i}(k)$ for the no-control case; $T = 60$ seconds.

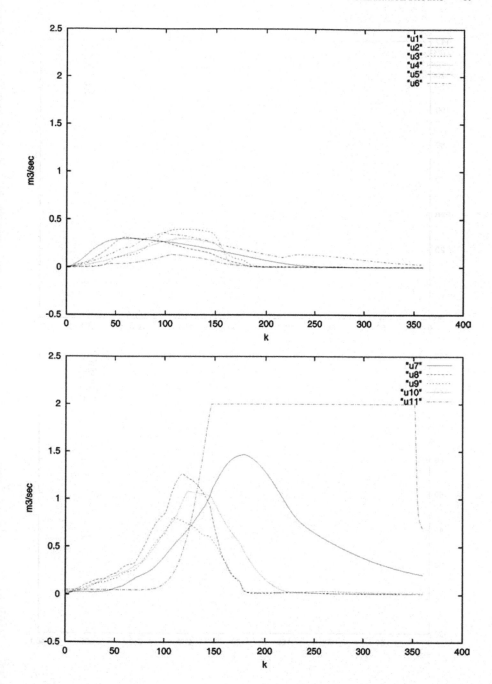

Figure 7.7. Scenario 3: Reservoir outflows $u_i(k)$ for the no-control case; $T = 60$ seconds.

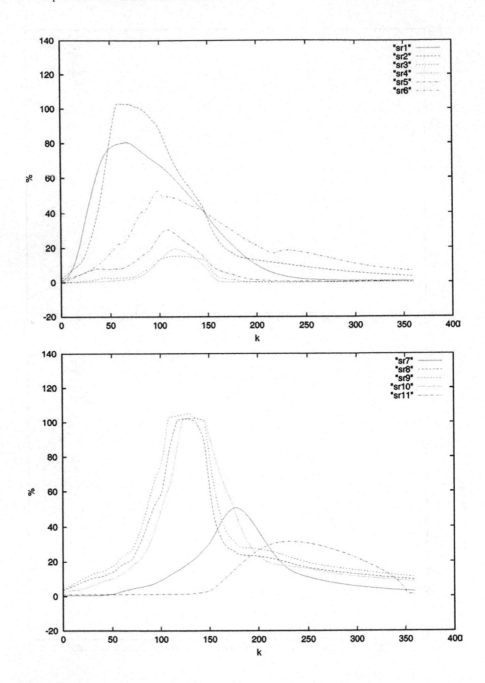

Figure 7.8. Scenario 3: Relative reservoir storages $(V_i(k)/V_{i,max})100\%$ for the no-control case; $T = 60$ seconds.

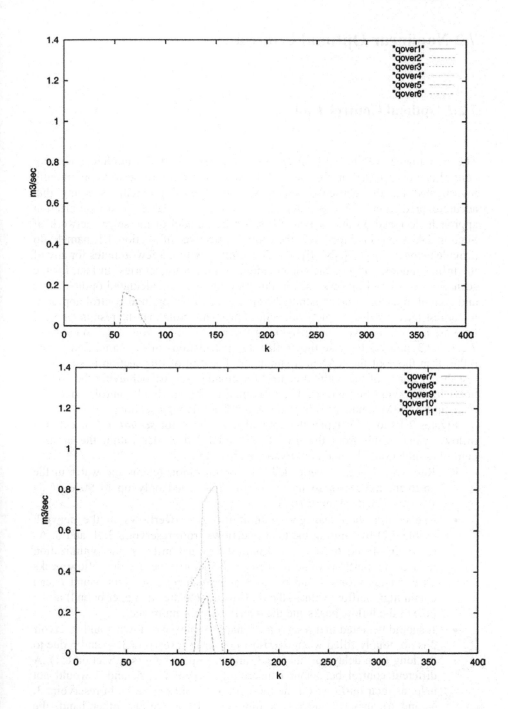

Figure 7.9. Scenario 3: Reservoir overflows $q_{over,i}(k)$ for the no-control case; $T = 60$ seconds.

7.2 Nonlinear Optimal Control

7.2.1 Optimal Control Tool

The performance of the optimal control tool, that is, of the nonlinear optimal control approach based on the simplified model of the sewer network, is initially investigated for the particular sewer network control problem assuming that accurate predictions for the external inflows are available. The optimization approach, designed on the basis of the simplified model of the sewer network of Section 2.4, is expected to satisfy the control objectives of Section 4.1, namely, to provide *automatically* highly efficient flow control within a few minutes for any of the infinite number of possible combinations of inflow trajectories. In fact, for the scenarios of external inflows used in this monograph, the calculated optimal state and control trajectories demonstrate the efficiency of the optimal control approach to address the central sewer network control problem. Note that the results reported here are not quantitatively comparable to the results of Sections 7.1 (no-control), 7.2.2, 7.2.3, and 7.3 because they reflect the calculations of the simplified model rather than those of the KANSIM simulator. The aim of this section is simply to test the adequacy of the optimal control formulation toward achieving the control goals for the simplified model. The relevance of the obtained controls under the realistic KANSIM simulator is investigated in the following sections.

Figures 7.10 to 7.13 depict the optimal trajectories for *scenario 1* of external inflows which result from the use of the simplified model within the optimal control formulation. The main observations are as follows:

- Reservoir 7 is not overloaded. The optimization retains the water in the upstream reservoirs so that reservoir 7 is filled only up to 90% of its storage capacity (Figure 7.11).
- The optimization manages to limit the total overflows in the network (Table 7.2) by limiting the total overflows from reservoirs 3, 4, and 6. As reservoirs 4 and 6 have the strongest external inflows, the optimization reduces the outflow value for reservoir 3 shortly before the inflow peaks arrive at reservoirs 4 and 6. As a consequence, reservoirs 4 and 6 can obtain high outflow values (the outflow u_6 activates its upper bound) at the time of the inflow peaks and the overflows are minimized.
- It should be noted that reservoirs 1 and 5 are not overflowing and reservoir 2 is not totally filled when overflows occur in reservoirs 3, 4, and 6 due to the long time delays of links 1, 3, and 2, respectively (see Section 6.1). A different control behaviour concerning reservoirs 1, 5, and 2 would not help, as their outflows (at the moment of the inflow peaks in reservoirs 3, 4, and 6) arrive at node 3 a long time later. On the other hand, the unavoidable overflows are fairly homogeneous in reservoirs 3, 4, and 6, that is, the optimal solution in fact distributes the unavoidable overflows in space and time as much as possible.

- The outflow of reservoir 11 does not reach the flow capacity of the treatment plant (Figure 7.10). This is due to reservoir 7, which does not send the upstream water downward at a sufficiently high rate as its outflow is restricted from $u_{max,7}(V_7(k),k)$.

Figures 7.14 to 7.16 depict the optimal trajectories for *scenario 2* of external inflows. The main observations are as follows:

- The optimization manages to completely avoid any overflows in the network (Table 7.2), notably from reservoirs 1, 2, and 8 to 10, which produced overflows in the no-control case (Table 7.1). More specifically, during the critical period reservoir 7 has a small outflow, thus permitting reservoirs 8, 9, and 10 to obtain high outflow values, and consequently overflows from reservoirs 8, 9, and 10 do not occur.
- The outflow of reservoir 11 reaches the flow capacity of the treatment plant (Figure 7.14) as soon as possible and holds this value until the end of the control operation, thus leading to a quick emptying of the network.

Figures 7.17 to 7.19 depict the optimal trajectories for *scenario 3* of external inflows. The main observations are as follows:

- The optimization manages to completely avoid any overflows in the network (Table 7.2), notably from reservoirs 2 and 8 to 10 that produced overflows in the no-control case (Table 7.1). The nonlinear optimal control, knowing about the inflow peaks that are going to arrive at reservoirs 8 and 9 at around 106 minutes and at reservoir 10 at around 114 minutes, reduces the outflow of reservoir 7 earlier, at around 30 minutes, and retains a small outflow value for reservoir 7 for another 60 minutes; thus reservoirs 8, 9, and 10 can obtain high outflows and consequently no overflows arise at these reservoirs.
- The outflow of reservoir 11 reaches the flow capacity of the treatment plant (Figure 7.17) as soon as possible. However, it does not keep this value fully until the end of the control operation as reservoir 7 does not send the upstream water downward at a sufficiently high rate (its outflow is restricted from $u_{max,7}(V_7(k),k)$ during the time period [144 minutes, 192 minutes]) and reservoirs 8 and 9 have small outflows or outflows equal to u_{min} during the time period [174 minutes, 360 minutes].

It should be noted that for all three scenarios:

- The flow capacity is not exceeded in any sewer stretch (Figures 7.13, 7.16, and 7.19).
- The desired distribution of storage volume is taken into account according to the priority of this subgoal. Thus, during the emptying phase, there is free storage space for a possible rain event in every reservoir (Figures 7.11, 7.15, and 7.18), see particularly for scenario 2 (Figure 7.15) for the time period [300 minutes, 360 minutes] and for scenario 3 (Figure 7.18) for the time period [318 minutes, 360 minutes].
- The term concerning abrupt changes of releases is considered to a satisfactory degree (Figures 7.10, 7.14, and 7.17).

Table 7.2. Reservoir overflows and overload of reservoir 7 in [m³] for nonlinear optimal control.

Reservoir	Scenario 1	Scenario 2	Scenario 3
1	0	0	0
2	0	0	0
3	392	0	0
4	529	0	0
5	0	0	0
6	381	0	0
8	0	0	0
9	0	0	0
10	0	0	0
11	0	0	0
Total	1302	0	0
7	0	0	0

7.2.2 Open-Loop Application

The performance of the nonlinear optimal control approach is now investigated in an open-loop manner in order to examine the adequacy of the control decisions for the particular sewer network control problem under more realistic simulation conditions. The optimal control trajectories that are derived from the optimal control tool of Section 7.2.1 are applied as reference trajectories for the local direct control to the simulation program KANSIM so as to enable an initial identification of the potential differences between the simplified and the accurate model of the sewer network.

The results obtained for the three scenarios of external inflows are summarized in Table 7.3, and Figures 7.20 to 7.23 depict the resulting trajectories for *scenario 3* of external inflows. The observations made here for scenario 3 apply similarly to the other two scenarios of external inflows. It can be seen from Figure 7.20 that the reservoir outflows for the open-loop application are quite similar to those of the nonlinear optimal control (Figure 7.17), that is, that for most time intervals the outflow of each reservoir is equal to the corresponding predefined reference value. On the other hand, the relative reservoir storages (Figure 7.21) and the reservoir overflows (Figure 7.22, Table 7.3) for the open-loop application differ from the ones of nonlinear optimal control (Figures 7.18, Table 7.2). These differences

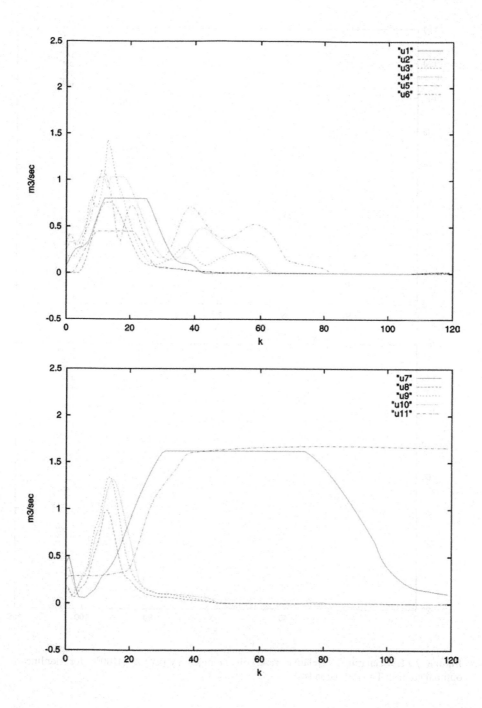

Figure 7.10. Scenario 1: Reservoir outflows $u_i(k)$ for nonlinear optimal control; T = 180 seconds.

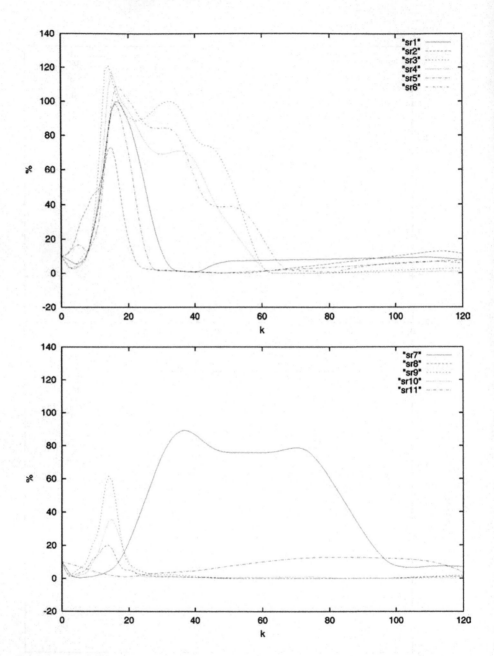

Figure 7.11. Scenario 1: Relative reservoir storages $(V_i(k)/V_{i,max})100\%$ for nonlinear optimal control; T = 180 seconds.

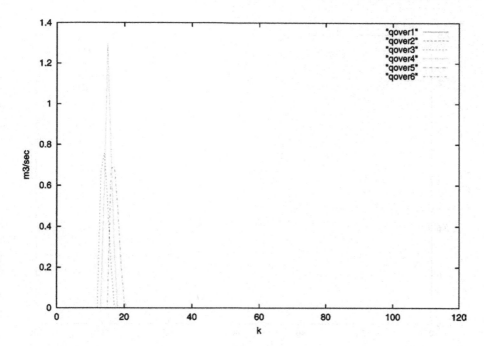

Figure 7.12. Scenario 1: Reservoir overflows $q_{over,i}(k)$ for nonlinear optimal control for $i=1,...6$; $T = 180$ seconds.

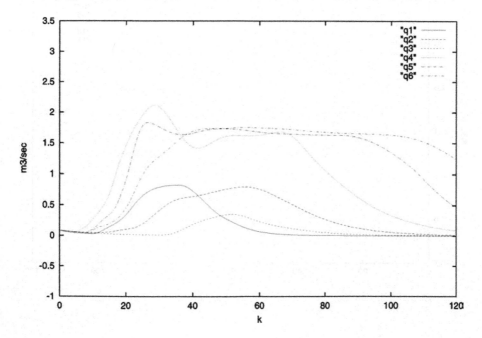

Figure 7.13. Scenario 1: Link outflows $q_i(k)$ for nonlinear optimal control; $T = 180$ seconds.

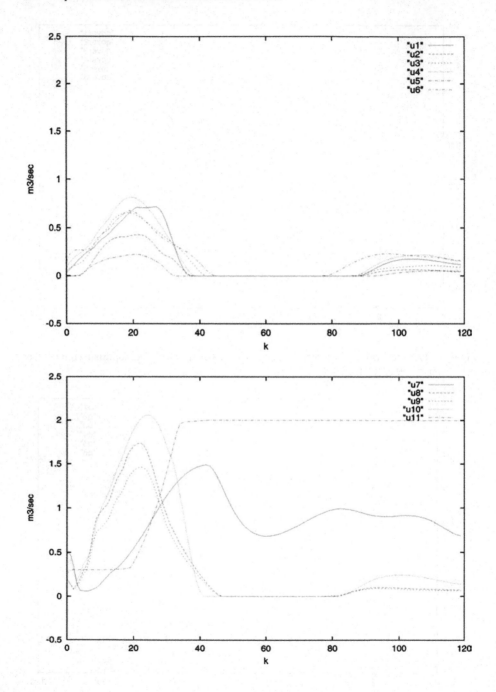

Figure 7.14. Scenario 2: Reservoir outflows $u_i(k)$ for nonlinear optimal control; $T = 180$ seconds.

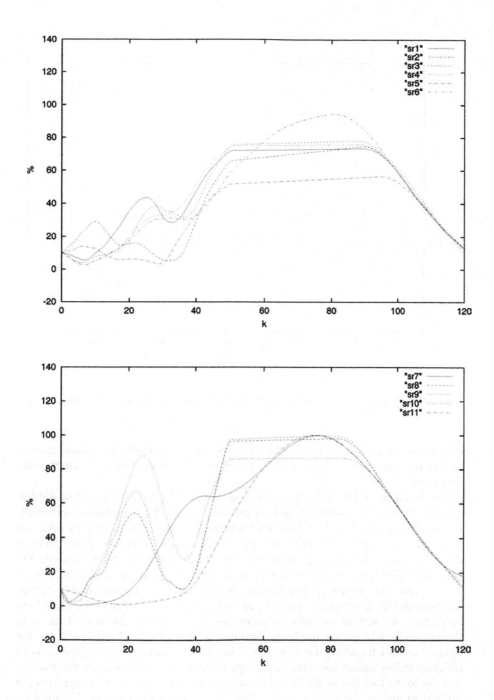

Figure 7.15. Scenario 2: Relative reservoir storages $(V_i(k)/V_{i,max})100\%$ for nonlinear optimal control; T = 180 seconds.

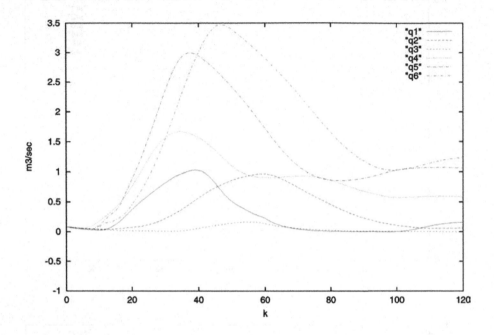

Figure 7.16. Scenario 2: Link outflows $q_i(k)$ for nonlinear optimal control; T = 180 seconds.

are due to the simplifications of the model used in the nonlinear optimal control and to the parameters of this model of the sewer network that were empirically estimated.

It should be noted that reservoirs 1 to 6 and 8 to 10 have equal or similar external inflows for nonlinear optimal control and for the open-loop application, whereas reservoirs 7 and 11 have different inflows (q_4 and q_6, respectively) for some time intervals, that is, the inflow peaks occur at different time steps and have different values. This fact, in addition to the model simplifications, leads to the significant difference in the storage of reservoir 7 between the nonlinear optimal control and the open-loop application, which leads to an undesirable strong overload of this reservoir. It may be seen from Figure 7.23 that in the open-loop application the peak of the inflow to reservoir 7 (q_4) is greater and arrives earlier at the reservoir than in the nonlinear optimal control (Figure 7.19). Thus, as the optimal control trajectory for reservoir 7 has been calculated taking into account different inflow values for some time steps than the ones appearing in the process, and due to the fact that in the open-loop application no real-time measurements are utilized to update the control trajectories, the overload of reservoir 7 cannot be avoided. This demonstrates the need for application of a rolling horizon procedure that utilizes real-time measurements to update the control trajectories at specific time periods according to Section 4.4. In fact it will be seen in Section 7.2.3, where

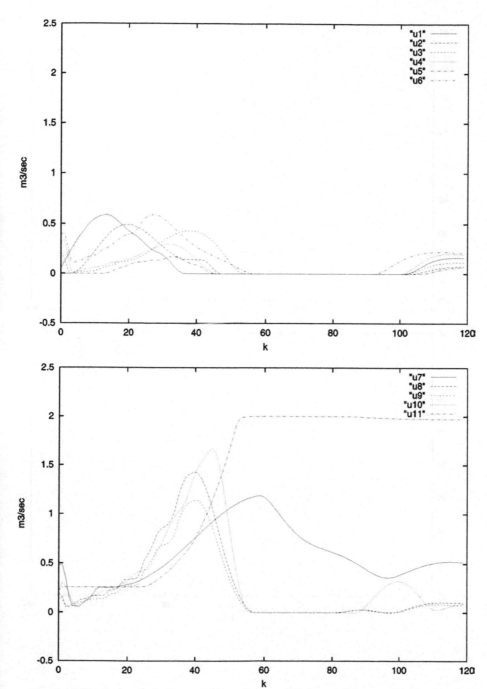

Figure 7.17. Scenario 3: Reservoir outflows $u_i(k)$ for nonlinear optimal control; T = 180 seconds.

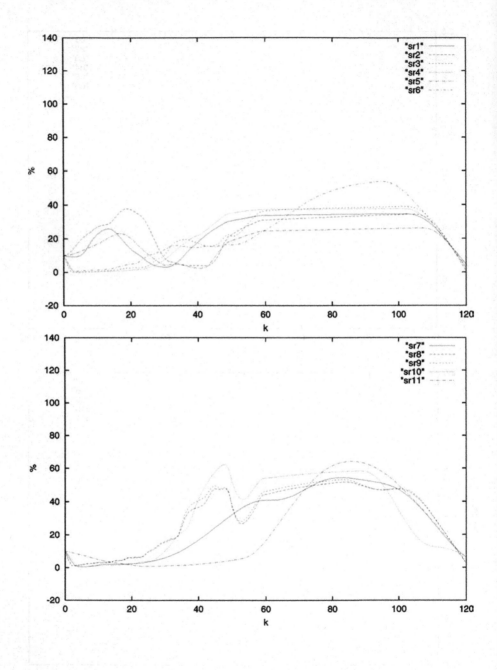

Figure 7.18. Scenario 3: Relative reservoir storages $(V_i(k)/V_{i,max})100\%$ for nonlinear optimal control; T = 180 seconds.

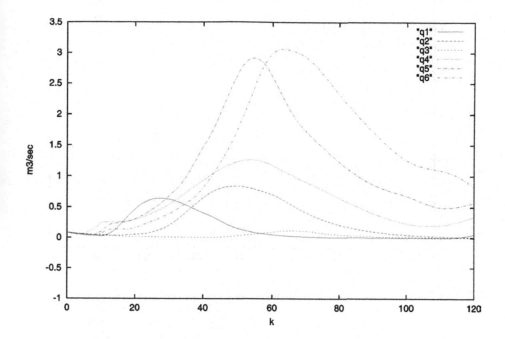

Figure 7.19. Scenario 3: Link outflows $q_i(k)$ for nonlinear optimal control; $T = 180$ seconds.

the rolling-horizon procedure is used, that the overload of reservoir 7 is quite small or zero for all three scenarios of external inflows.

Table 7.3. Reservoir overflows and overload of reservoir 7 in $[m^3]$ for nonlinear optimal control (open-loop).

Reservoir	Scenario 1	Scenario 2	Scenario 3
1	112	0	0
2	31	53	0
3	176	0	0
4	0	0	0
5	268	0	0
6	16	0	0
8	0	282	0
9	0	603	239
10	0	387	0
11	0	0	0
Total	**603**	**1325**	**239**
7	**2712**	**1202**	**138**

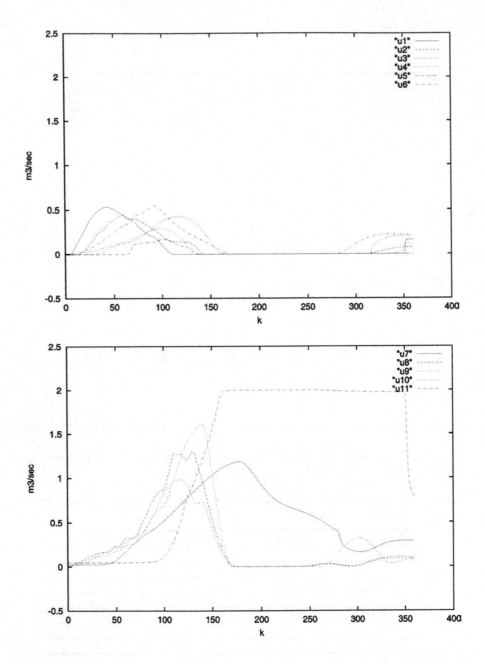

Figure 7.20. Scenario 3: Reservoir outflows $u_i(k)$ for nonlinear optimal control (open-loop); T = 60 seconds.

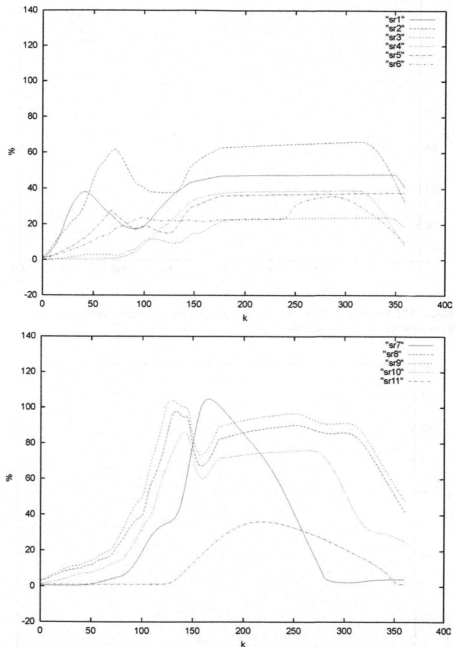

Figure 7.21. Scenario 3: Relative reservoir storages $(V_i(k)/V_{i,max})100\%$ for nonlinear optimal control (open-loop); $T = 60$ seconds.

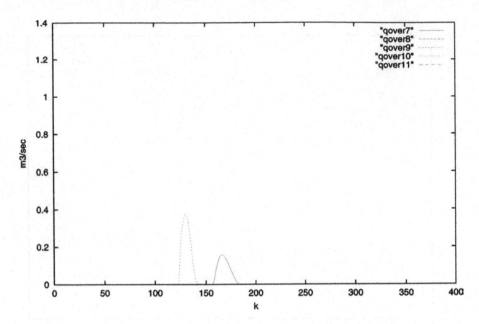

Figure 7.22. Scenario 3: Reservoir overflows $q_{over,i}(k)$ for nonlinear optimal control (open-loop) for i=7,…11; T = 60 seconds.

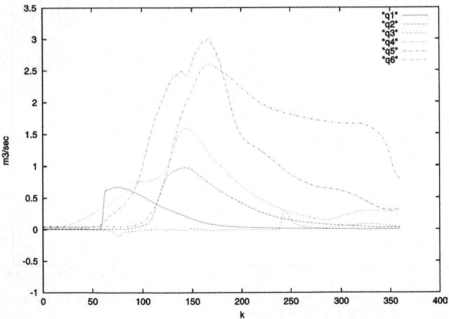

Figure 7.23. Scenario 3: Link outflows qi(k) for nonlinear optimal control (open-loop); T = 60 seconds.

7.2.3 Rolling Horizon Application

7.2.3.1 Investigated Cases

The efficiency of the nonlinear optimal control is tested using a closed-loop control structure – *optimization with rolling horizon* (Section 4.4). [The on-line application of nonlinear optimal control is possible, as optimization needs a few minutes to reach the minimum (Section 4.3.7)]. Using the rolling horizon, the impact of inaccurate predictions, modelling inaccuracies and unexpected disturbances remains limited (Marinaki and Papageorgiou, 2001). The disadvantages of the open-loop application (Section 7.2.2), where no real process measurements are utilized, are largely eliminated when the closed-loop application is employed and thus the control system efficiency is significantly improved compared to the open-loop application.

For the particular sewer network control problem, different repetition periods k_R and different optimization horizons K for each optimization run are used, in order to investigate the impact of k_R and K on the control performance. It is expected that the results should tend to be better for smaller values of k_R (due to more frequent updating, which eliminates the impact of past modelling and prediction errors) and for greater values of K (due to less myopic control). It should be noted, however, that when the rolling horizon with incomplete inflow information is applied, the results might not always get better for greater values of K as the accuracy of the predictions deteriorates. The optimization is repeated every k_R= 1, 2, 3, or 4 time instants. As the discrete time interval is T = 180 seconds for the control and T = 60 seconds for the simulation, a repetition period of k_R = 1 or 2 or 3 or 4 means that the optimization is repeated every k = 0, 3, 6, ... or k = 0, 6, 12, ... or k = 0, 9, 18,... or k = 0, 12, 24, ...time steps of the simulation, respectively. The investigated optimization horizons K correspond to time horizons of 1 h, 2 h, 2.5 h, 3 h, 3.5 h, and 4 h. Note that in the following subsections we may express the time horizon K either in time units (h) or in number of time steps, hopefully without creating confusion.

7.2.3.2 Rolling Horizon with Complete Inflow Information

In this section, it is assumed that accurate inflow predictions are available for the whole optimization run, that is, $K_p = K$ (Section 4.4). Although this assumption is rather unrealistic for the real application, it delivers an upper bound of system performance, which is useful for the assessment of the impact of inaccurate inflow predictions. The comparison between the different cases is made on the basis of total reservoir overflows and of the overload of reservoir 7, as the main task of the control system is the minimization of overflows and the avoidance of overload in storage elements, which do not have overflows.

The results obtained for the three scenarios of external inflows are summarized in Tables 7.4 to 7.6. These tables demonstrate the following:

- For *scenario 1*, in all cases the total reservoir overflows are strongly reduced, compared to the results of no-control, and, in most cases, are smaller than the results of the open-loop application. Moreover, as k_R is increased, the total overflows usually increase for all the optimization horizons, whereas as K is increased, the total overflows usually decrease for all repetition periods. The overload of reservoir 7 is significantly smaller than in the open-loop application, and, in most cases, smaller than in the no-control case. The most satisfactory results with respect to reservoir overflows and overload of reservoir 7 are taken when $k_R = 1$ and K = 4 hours. However, as there are no significant deviations between the results, we can say that the control results for all combinations of K and k_R are very satisfactory.

- For *scenario 2*, in all cases the total reservoir overflows are significantly reduced compared to the no-control case and are quite smaller compared to the open-loop application. Moreover, as k_R is increased, the total overflows usually increase for all optimization horizons, and as K is increased, the total overflows usually decrease for all repetition periods. Reservoir 7 has a small overload in most cases (this overload is significantly smaller than the one in the open-loop application). Taking into account both the reservoir overflows and the overload of reservoir 7 and the fact that the deviations between the results are relatively small, we can say that for all combinations of K and k_R, the control behaviour is very efficient.

- For *scenario 3*, in all cases the total reservoir overflows are much less than the ones of the no-control case and of the open-loop application. The overload of reservoir 7 is equal to zero in all cases and, thus, it is smaller than in the open-loop application. Taking into account both the reservoir overflows and the overload of reservoir 7, we can say that for all combinations of K and k_R the control results are very satisfactory.

These results demonstrate the efficiency of the rolling horizon with complete inflow information in solving the sewer network control problem.

The outflow, overflow, and storage trajectories of *scenario 1* when $k_R = 1$ and K = 2 hours are commented on the following discussion in order to analyse in more detail the behaviour of nonlinear optimal control when applied in a closed-loop control structure with complete inflow information. The resulting trajectories for $k_R = 1$ and K = 2 hours are presented in Figures 7.24 to 7.26 and Table 7.16 displays the corresponding overflow and overload values. From these results we can draw the following conclusions:

- Optimal control reduces or avoids overflows from reservoirs 1 to 6 and 9 (Table 7.16) that were overflowing in the no-control case (Tables 7.1) and in the open-loop application (Table 7.3). This is achieved by increasing the early outflows from these reservoirs in a way that exploits the flow capacity of the network links without strongly overloading reservoir 7 (its overload is decreased compared to the no-control case and to the open-loop application). Eventually, the reservoir outflows are changed as appropriate to minimize the unavoidable overflows.

- The inflow of the treatment plant $u_{11} = r$, never reaches its flow capacity of 2 m^3/s due to the limited capacity of reservoir 7 outflow, which is fully utilized during $80 \leq k \leq 220$.
- Despite the inaccuracies of the simplified model, the optimization manages to deliver reasonable results that automatically and intelligently take into account and exploit the network structure, the long time delays in network links, the time-variation of the inflows, and the available reservoir capacity so as to lead to excellent control performance in a highly complex problem environment.

Table 7.4. Scenario 1: Total reservoir overflows and overload of reservoir 7 in [m^3] for nonlinear optimal control (closed-loop) for $K_p = K$.

		1h	2h	2.5h	3h	3.5h	4h
$k_R = 1$	Overflows	658	553	532	506	503	412
	Overload	83	14	70	16	15	70
$k_R = 2$	Overflows	710	554	534	519	509	422
	Overload	70	42	37	64	14	71
$k_R = 3$	Overflows	725	543	524	549	513	474
	Overload	174	117	62	105	88	111
$k_R = 4$	Overflows	710	545	536	556	573	576
	Overload	94	101	115	125	107	102

Table 7.5. Scenario 2: Total reservoir overflows and overload of reservoir 7 in [m^3] for nonlinear optimal control (closed-loop) for $K_p = K$.

		1h	2h	2.5h	3h	3.5h	4h
$k_R = 1$	Overflows	439	272	274	263	248	241
	Overload	49	32	33	25	26	26
$k_R = 2$	Overflows	484	334	281	284	277	247
	Overload	21	29	20	48	28	17
$k_R = 3$	Overflows	475	360	282	277	281	244
	Overload	49	53	12	15	29	16
$k_R = 4$	Overflows	514	395	285	280	287	266
	Overload	61	92	60	20	30	53

Table 7.6. Scenario 3: Total reservoir overflows and overload of reservoir 7 in [m^3] for nonlinear optimal control (closed-loop) for $K_p = K$.

		1h	2h	2.5h	3h	3.5h	4h
$k_R = 1$	Overflows	175	177	175	172	170	159
	Overload	0	0	0	0	0	0
$k_R = 2$	Overflows	176	178	176	171	171	167
	Overload	0	0	0	0	0	0
$k_R = 3$	Overflows	176	176	175	174	170	169
	Overload	0	0	0	0	0	0
$k_R = 4$	Overflows	176	176	175	171	169	167
	Overload	0	0	0	0	0	0

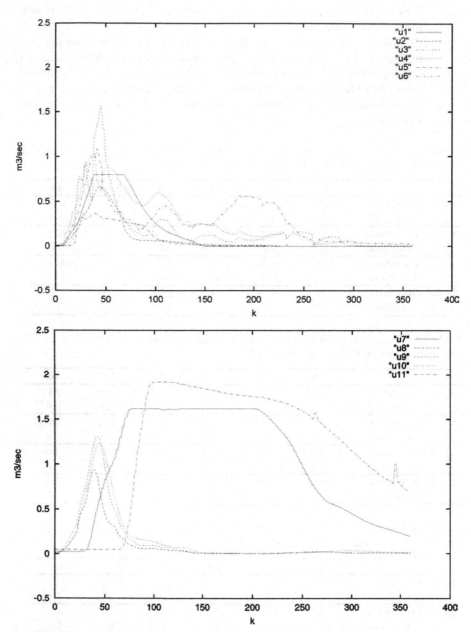

Figure 7.24. Scenario 1: Reservoir outflows $u_i(k)$ for nonlinear optimal control (closed-loop), $K_p = K$, $k_R = 1$, $K = 2$ hours; $T = 60$ seconds.

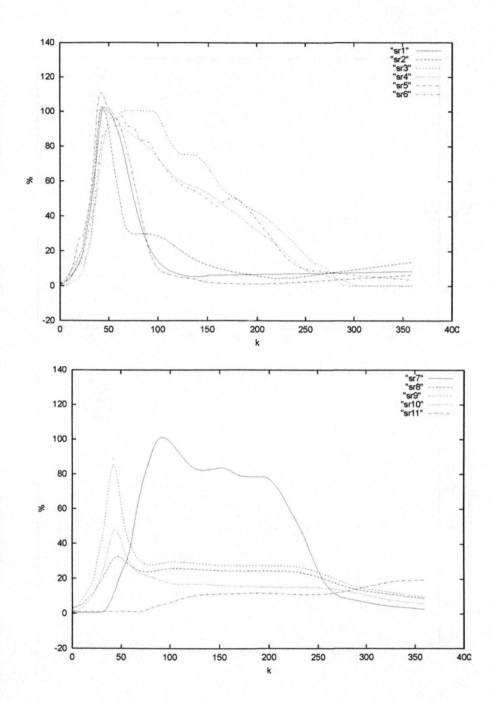

Figure 7.25. Scenario 1: Relative reservoir storages $(V_i(k)/V_{i,max})100\%$ for nonlinear optimal control (closed-loop), $K_p = K$, $k_R = 1$, $K = 2$ hours; $T = 60$ seconds.

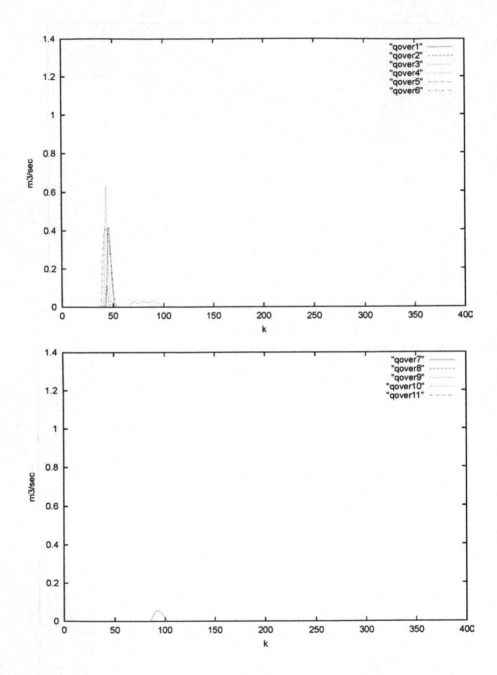

Figure 7.26. Scenario 1: Reservoir overflows $q_{over,i}(k)$ for nonlinear optimal control (closed-loop), $K_p = K$, $k_R = 1$, $K = 2$ hours; $T = 60$ seconds.

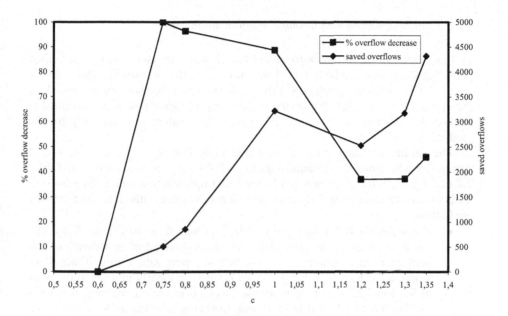

Figure 7.27. Percent overflow decrease and saved overflows (in m^3) for different inflow charge.

There is a general belief and expectation that the level of overflow reduction achieved by efficient sewer network control as compared to the no-control case depends on the level of external inflows. More specifically, if the inflows are low, there is no overflow occurring in the no-control case, hence there is hardly anything to improve by control application. When external inflows are relatively high, leading to according overflows in some reservoirs, optimal control may avoid or reduce the overflows via utilization of reserve capacity in other reservoirs. Finally, if the inflows are very high, leading to generalized overflows in virtually all reservoirs, there is little potential for improvement via network control due to the lack of reserve capacity. In summary, highest overflow reductions are expected for medium-level inflows.

To investigate this issue, the closed-loop control structure with $k_R = 1$ and $K = 4$ hours has been applied to scenario-1 inflows multiplied by a factor c. The corresponding overflow results are depicted in Figure 7.27. For $c \leq 0.6$ the total overflows without control are zero, hence there is nothing to decrease. For $c = 0.75$, some overflow appears in the no-control case, whereas the optimal control manages to completely avoid overflow, hence a reduction of 100% is achieved. For $c \geq 0.8$ the degree of amelioration due to optimal control is monotonically decreasing and for $c \geq 1.2$ the reduction of overflows is less than 50%. However, the saved overflows [the total overflows in the no-control case minus the total overflows in the optimal control (in m³)] increase for values of c ≤ 1. These tests confirm the general belief mentioned in the previous paragraph.

7.2.3.3 Rolling Horizon with Incomplete Inflow Information

In this section, the rolling-horizon optimal control is used again, but inflow predictions are only available for 30 minutes (K_p = 10), 60 minutes, (K_p = 20) or not at all (K_p = 0). The calculation of the predictions over the horizon K, needed for the optimization runs, is made by the use of the prolongation scheme described in Section 4.4, whereby in the third case, past inflow values are used only for the prediction.

The results obtained using a repetition period of k_R = 1, 2, 3, or 4 and optimization horizons K = 2 hours, 3 hours, and 4 hours, are summarized in Tables 7.7 to 7.15. From these tables it can be seen that optimization with rolling horizon leads to very reasonable and efficient control results even without accurate inflow predictions. Thus:

- For *scenario 1*, for K_p = 20 (Table 7.7) and K_p = 10 (Table 7.10) the results obtained are quite similar to the ones obtained when accurate inflow predictions are available for the whole optimization run (Table 7.4), whereas for K_p = 0 (Table 7.13) the results are clearly inferior to the ones of Section 7.2.3.2. For all three cases, as k_R is increased, the total overflows usually increase for all optimization horizons, whereas as K is increased, the total overflows usually decrease for all repetition periods. The overload of reservoir 7 is in almost all cases significantly smaller than the one in the no-control case. Generally, for all three cases and taking into account both the total overflows and the overload of reservoir 7, the results are significantly better than the ones of the no-control case.

- For *scenario 2*, for K_p = 20 (Table 7.8) and K_p = 10 (Table 7.11) the results obtained are quite similar and in some cases slightly inferior to the ones obtained when accurate inflow predictions are available for the whole optimization run (Table 7.5), whereas for K_p = 0 (Table 7.14) the results slightly deteriorate compared to the ones of Section 7.2.3.2. For all combinations of K and k_R, the control results are very efficient compared to the no-control case.

- For *scenario 3*, for K_p = 20 (Table 7.9) the results obtained are quite similar to the ones obtained when accurate inflow predictions are available for the whole optimization run (Table 7.6); for K_p = 10 (Table 7.12) the results are quite similar in most cases (when k_R = 3 and k_R = 4 for all optimization horizons the results slightly deteriorate compared to the ones of Section 7.2.3.2); whereas for K_p = 0 (Table 7.15) the results are significantly inferior to the ones of Section 7.2.3.2. These significant differences between the case K_p=0 and the rolling horizon with complete inflow information are due to the form of this inflow scenario. As this scenario is temporally inhomogeneous, an underestimation or overestimation of the future inflow values from the prolongation scheme used can lead to quite significant deviations from the case of rolling horizon with complete inflow information. However, even these results are

Table 7.7. Scenario 1: Total reservoir overflows and overload of reservoir 7 in [m^3] for nonlinear optimal control (closed-loop) for K_p=20.

		2h	3h	4h
k_R=1	Overflows	509	502	447
	Overload	35	31	8
k_R=2	Overflows	511	522	459
	Overload	18	27	25
k_R=3	Overflows	518	575	461
	Overload	18	65	69
k_R=4	Overflows	501	556	463
	Overload	102	138	108

Table 7.8. Scenario 2: Total reservoir overflows and overload of reservoir 7 in [m^3] for nonlinear optimal control (closed-loop) for K_p=20.

		2h	3h	4h
k_R=1	Overflows	431	350	241
	Overload	45	40	29
k_R=2	Overflows	505	462	441
	Overload	40	37	22
k_R=3	Overflows	480	485	491
	Overload	48	43	25
k_R=4	Overflows	703	451	467
	Overload	21	59	60

Table 7.9. Scenario 3: Total reservoir overflows and overload of reservoir 7 in [m^3] for nonlinear optimal control (closed-loop) for K_p=20.

		2h	3h	4h
k_R=1	Overflows	169	170	164
	Overload	0	0	0
k_R=2	Overflows	160	172	175
	Overload	4	0	0
k_R=3	Overflows	164	171	170
	Overload	0	0	0
k_R=4	Overflows	168	172	173
	Overload	4	0	5

quite satisfactory as in almost all cases and for all combination of K and k_R, a significant reduction in the total reservoir overflows is observed compared to the no-control case.

To analyse in more detail the behaviour of nonlinear optimal control when applied in a closed-loop control structure with incomplete inflow information, the results obtained for one specific case (for scenario 1 and for k_R = 1 and K = 2 hours) when K_p is equal to K, 20, 10, or 0, are presented in Table 7.16. These results are almost identical for K_p = K, 20, or 10 while a significant difference from the case K_p = K appears only when K_p = 0. The optimal trajectories for this latter case are presented in Figures 7.28 to 7.30 and are quite similar to the ones obtained with complete inflow information (Figures 7.24 to 7.26). This indicates that the behaviour of rolling horizon with incomplete inflow information is very satisfactory and is quite similar to the one presented in Section 7.2.3.2 even if no inflow predictions are available.

Table 7.10. Scenario 1: Total reservoir overflows and overload of reservoir 7 in [m³] for nonlinear optimal control (closed-loop) for K_p=10.

		2h	3h	4h
k_R=1	Overflows	522	524	471
	Overload	122	79	92
k_R=2	Overflows	530	538	502
	Overload	92	75	60
k_R=3	Overflows	560	580	550
	Overload	127	84	92
k_R=4	Overflows	592	623	544
	Overload	88	88	106

Table 7.11. Scenario 2: Total reservoir overflows and overload of reservoir 7 in [m³] for nonlinear optimal control (closed-loop) for K_p=10.

		2h	3h	4h
k_R=1	Overflows	358	363	407
	Overload	10	29	27
k_R=2	Overflows	425	408	410
	Overload	35	23	21
k_R=3	Overflows	448	354	450
	Overload	17	21	12
k_R=4	Overflows	528	536	580
	Overload	22	25	32

Table 7.12. Scenario 3: Total reservoir overflows and overload of reservoir 7 in [m³] for nonlinear optimal control (closed-loop) for K_p=10.

		2h	3h	4h
k_R=1	Overflows	170	171	164
	Overload	0	3	4
k_R=2	Overflows	178	187	182
	Overload	10	15	10
k_R=3	Overflows	215	225	240
	Overload	20	21	28
k_R=4	Overflows	250	270	290
	Overload	40	34	41

Table 7.13. Scenario 1: Total reservoir overflows and overload of reservoir 7 in [m³] for nonlinear optimal control (closed-loop) for K_p=0.

		2h	3h	4h
k_R=1	Overflows	1118	1005	882
	Overload	0	0	24
k_R=2	Overflows	1120	932	846
	Overload	0	10	29
k_R=3	Overflows	1251	1289	1204
	Overload	20	45	31
k_R=4	Overflows	2436	2028	1986
	Overload	0	34	89

Table 7.14. Scenario 2: Total reservoir overflows and overload of reservoir 7 in [m³] for nonlinear optimal control (closed-loop) for K_p=0.

		2h	3h	4h
k_R=1	Overflows	517	516	513
	Overload	0	0	0
k_R=2	Overflows	504	501	510
	Overload	0	0	0
k_R=3	Overflows	556	554	565
	Overload	0	0	0
k_R=4	Overflows	685	686	697
	Overload	0	0	0

Table 7.15. Scenario 3: Total reservoir overflows and overload of reservoir 7 in [m³] for nonlinear optimal control (closed-loop) for K_p=0.

		2h	3h	4h
k_R=1	Overflows	724	730	740
	Overload	0	0	0
k_R=2	Overflows	833	839	821
	Overload	0	0	0
k_R=3	Overflows	847	840	851
	Overload	0	0	0
k_R=4	Overflows	855	860	865
	Overload	0	0	0

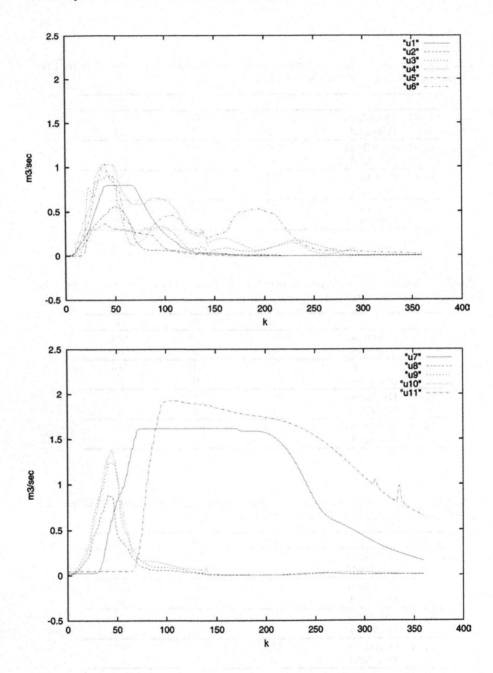

Figure 7.28. Scenario 1: Reservoir outflows $u_i(k)$ for nonlinear optimal control (closed-loop), $K_p = 0$, $k_R = 1$, $K = 2$ hours; $T = 60$ seconds.

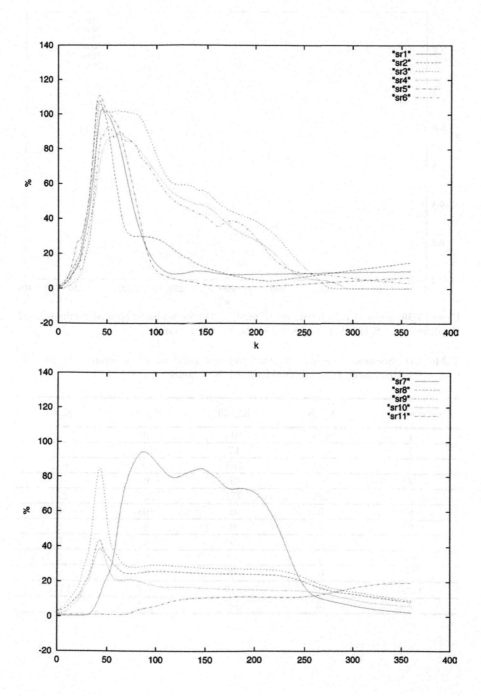

Figure 7.29. Scenario 1: Relative reservoir storages $(V_i(k)/V_{i,max})100\%$ for nonlinear optimal control (closed-loop), $K_p = 0$, $k_R = 1$, $K = 2$ hours; $T = 60$ seconds.

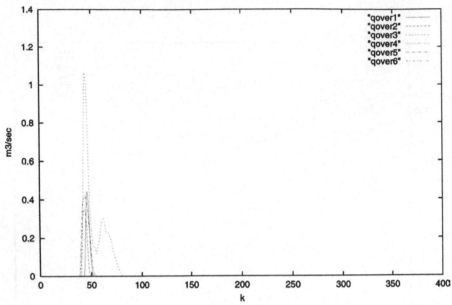

Figure 7.30. Scenario 1: Reservoir overflows $q_{over,i}(k)$ for nonlinear optimal control (closed-loop) for i=1, .., 6, $K_p = 0$, $k_R = 1$, K = 2 hours; T = 60 seconds.

Table 7.16. Scenario 1: Reservoir overflows and overload of reservoir 7 in $[m^3]$ for nonlinear optimal control (closed-loop) for k_R=1, K=2 hours.

Reservoir	K_p=K	K_p=20	K_p=10	K_p=0
1	108	108	108	115
2	17	17	22	119
3	163	164	65	662
4	0	0	0	0
5	265	221	221	220
6	0	0	108	0
8	0	0	0	0
9	0	0	0	0
10	0	0	0	0
11	0	0	0	0
Total	553	510	524	1116
7	14	35	122	0

7.2.4 General Observations

The nonlinear optimal control, designed on the basis of a simplified model of the sewer network, provides automatically highly efficient control results for different scenarios of external inflows. However, the performance of the nonlinear optimal control approach when applied in an open-loop manner to the particular sewer network control problem was not satisfactory (a reduction of total reservoir overflows was achieved but the overload of reservoir 7 was strongly increased compared to the no-control case). This behaviour is due to the simplifications of the model of the sewer network used in the nonlinear optimal control approach and to the fact that no real process measurements are utilized in the open-loop control strategy. When the nonlinear optimal control was embedded in a closed-loop control structure with updated inflow predictions and updated initial conditions, the control results were very satisfactory.

In the rolling horizon, different repetition periods k_R and different optimization horizons K for each optimization run were investigated, each time with inflow predictions of different accuracy levels. The results obtained were better when accurate inflow predictions were used, but, even with inaccurate or missing inflow predictions, the obtained control results are very satisfactory and in all cases significantly better than the ones of the no-control case. The results obtained for all combinations of K and k_R, both with complete and incomplete inflow information, were very satisfactory and did not lead to myopic control behaviour. Generally, we can say that the results of the optimization with rolling horizon were very efficient and were significantly better than the ones of the no-control case even without inflow predictions.

7.3 Multivariable Regulator

7.3.1 Multivariable Regulator without Feedforward Terms

The multivariable feedback controller is reducing overflows from the network indirectly, by means of homogeneous storage distribution (Section 5.2). It does not explicitly consider the control constraints, which are imposed (where necessary) heuristically after calculation of the feedback law or are imposed by the physical laws used in the simulation (e.g., link flow capacities). In addition, the regulator without feedforward terms merely **reacts** to the impact of inflows on the measurable storages as mentioned in Section 5.2 without the on-line use of a model and without any consideration of inflow predictions. These main observations about the multivariable regulator's behaviour are visible in the control results obtained for the three scenarios of external inflows used. Despite these simplifications, the obtained control and state trajectories demonstrate the efficiency of the feedback controller to solve the central sewer network control problem.

It should be noted that in all scenarios of external inflows and for both multivariable regulators (with and without feedforward terms) the relative storage of reservoir 7 ($sr_7(k) = (V_7(k)/V_{7,max})100\%$) is not equalized with the relative storage of the other reservoirs ($sr_i(k) = (V_i(k)/V_{i,max})100\%$ for i=1,...,11, i≠7). This is mainly due to the fact that in this study the multivariable regulators are designed in order to equalize the relative storages of the reservoirs that are defined as $s_i(k) = ((V_i(k) - V_i^N)/(V_{i,max} - V_i^N))100\%$. Indeed, for this definition of the relative reservoir storage the equalization is very good; see Figure 7.33 for scenario 1. However, for this scenario during the period $55 \leq k \leq 217$ the simulation cannot take the outflow value of reservoir 7 ordered by the controller due to the existing flow capacity limit of link 5, thus leading to a relative storage of reservoir 7 (Figure 7.33), which is not perfectly equal to the storage of the other reservoirs during that period. [The trajectories $sr_i(k)$ rather than $s_i(k)$ are presented in the following discussion in order to enable comparison of the results of the multivariable regulators with the ones of the no-control case and of the nonlinear optimal control.]

Figures 7.31 to 7.34 depict the regulator results for *scenario 1* of external inflows. The main observations are as follows:

- As already mentioned in Section 7.1, reservoirs 1-6 and reservoir 9 overflow while a small overload is created in reservoir 7 in the no-control case. The multivariable regulator attempting to equalize the relative reservoir storages orders quite large outflows at almost all the reservoirs upstream of reservoir 7 (Figure 7.31) and finally succeeds in establishing a homogeneous distribution of most relative reservoir storages at some 100% of their storage capacity (Figure 7.32) during the critical period where overflows occur. An equalization at a higher storage level is not possible for reservoirs 8,...,11 because these reservoirs have small external inflows and hence small storages. However, around k = 80 where most reservoirs upstream of reservoir 7 have their relative reservoir storages around 60%, the regulator closes reservoirs 8 and 10 and increases the outflows of reservoirs 9 and 7 (the values of $u_7(k)$ are less than the ones ordered by the regulator, as mentioned previously) in order to achieve an equalization of the relative reservoir storages of 7 to 10 at this percentage. Generally, for this scenario, the multivariable regulator manages to significantly reduce the total reservoir overflows compared to the ones of the no-control case and to avoid the overload of reservoir 7 (Figure 7.34, Table 7.17).

- The treatment plant is fed with quite high flows, which helps the network to be emptied quite soon in order to have free storage space for a possible future rainfall.

Figures 7.35 to 7.37 depict the regulator results for the *scenario 2* of external inflows. The main observations are as follows:

- The multivariable regulator manages to significantly reduce the total overflows in the network (Figure 7.37, Table 7.17) compared to the no-control case. During the critical period where overflows occur, the regulator closes reservoir 7 in order to equalize its relative storage with that of the other reservoirs (Figure 7.36), and so reservoirs 8, 9, and 10, which are strongly overflowing in the no-control case, can have high outflows which leads to a significant reduction of their overflows.

- As Figure 7.35 indicates, the treatment plant is fed with its flow capacity, and so the network is emptied as soon as possible in order to have free storage space for a possible future rainfall.

Figures 7.38 to 7.40 depict the regulator results for the *scenario 3* of external inflows. The main observations are as follows:

- The multivariable regulator closes reservoir 7 in the beginning of the control period and orders small outflow for this reservoir afterward in order to equalize its relative storage with that of the other reservoirs (Figure 7.39). This permits reservoirs 8, 9, and 10 to have high outflows, which subsequently leads to lower overflows (Figure 7.40, Table 7.17) compared to those occurring in the no-control case.
- As Figure 7.38 shows, the inflow capacity of the treatment plant is fully used, and so there is free storage space for a possible future rainfall.

7.3.2 Multivariable Regulator with Feedforward Terms

The multivariable regulator with additional feedforward terms anticipates to some extent the impact of future inflows (Section 5.4). For the three scenarios of external inflows used, control results obtained by the multivariable regulator with additional feedforward terms are quite satisfactory. The results obtained using the multivariable regulator with feedforward terms are equally efficient or slightly superior to the control results obtained using the multivariable regulator without feedforward terms, depending on the particular inflow event.

Table 7.17. Reservoir overflows and overload of reservoir 7 in [m³] for multivariable regulator without feedforward terms.

Reservoir	Scenario 1	Scenario 2	Scenario 3
1	95	0	0
2	194	0	0
3	643	0	0
4	0	0	0
5	220	0	0
6	5	0	0
8	0	0	14
9	0	252	511
10	0	98	0
11	0	0	0
Total	1157	350	525
7	0	0	0

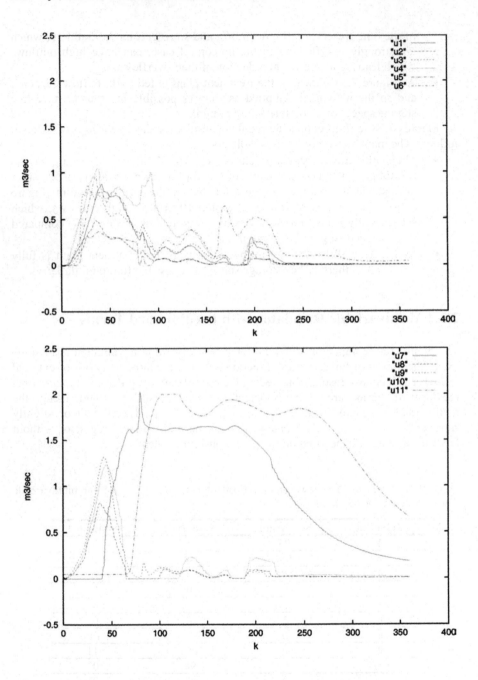

Figure 7.31. Scenario 1: Reservoir outflows $u_i(k)$ for multivariable regulator without feedforward terms; T = 60 seconds.

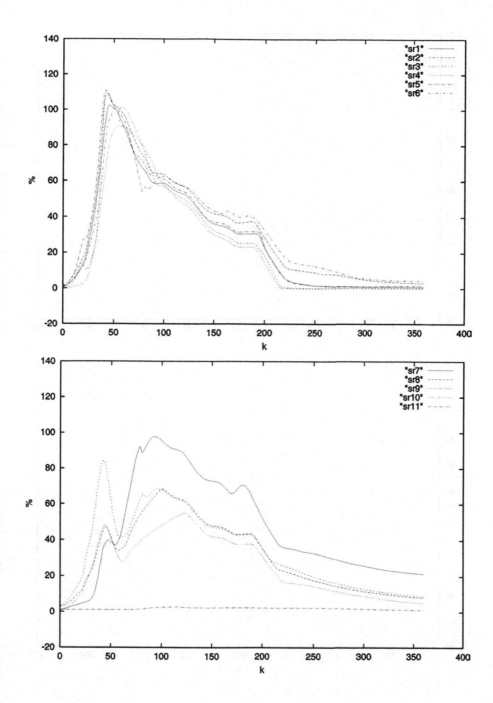

Figure 7.32. Scenario 1: Relative reservoir storages $(V_i(k)/V_{i,max})100\%$ for multivariable regulator without feedforward terms; $T = 60$ seconds.

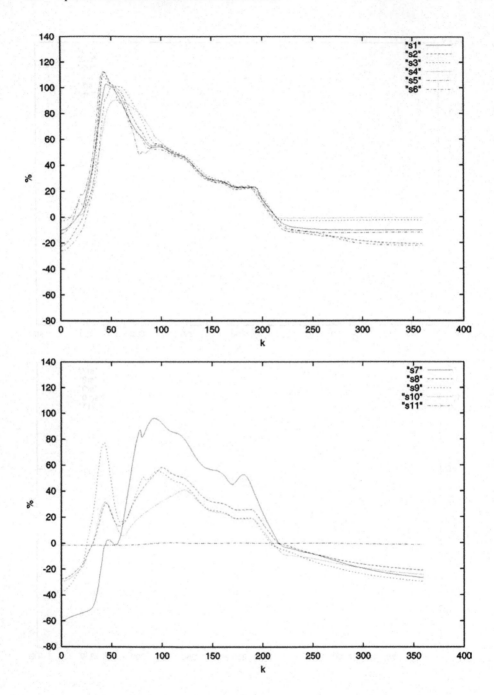

Figure 7.33. Scenario 1: Relative reservoir storages $((V_i(k) - V_i^N)/(V_{i,max} - V_i^N))100\%$ for multivariable regulator without feedforward terms; $T = 60$ seconds.

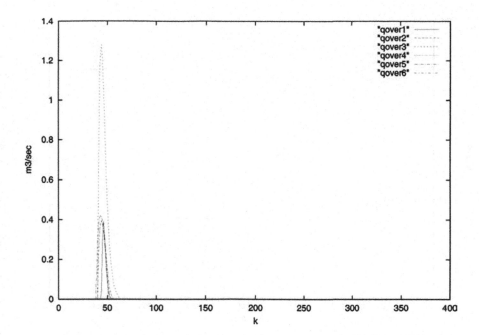

Figure 7.34. Scenario 1: Reservoir overflows $q_{over,i}(k)$ for multivariable regulator without feedforward terms for $i=1,...6$; $T = 60$ seconds.

7.3.2.1 Multivariable Regulator with Feedforward Terms and Accurate Inflow Predictions

When accurate inflow predictions are assumed available for the whole prediction horizon K_s (Section 5.4), the results of Table 7.18 and of Figures 7.41 to 7.43 for *scenario 1*, of Figures 7.44 to 7.46 for *scenario 2*, and of Figures 7.47 to 7.49 for *scenario 3* are obtained. These results indicate that the regulator with feedforward terms is very efficient.

For scenario 1 of external inflows the results obtained from the multivariable regulator with and without feedforward terms are almost equivalent. The control trajectories of multivariable regulator with feedforward terms (Figures 7.41 to 7.43) are quite similar for almost all reservoirs to the ones of the multivariable regulator without feedforward terms (Figures 7.31 to 7.34). However, the multivariable regulator with feedforward terms, knowing about the large inflow in reservoir 3, gives smaller outflows for reservoirs 1 and 2, and larger overflows for these reservoirs (Table 7.18) than the regulator without feedforward terms. By doing this, reservoir 3 can have larger outflow than with the regulator without feedforward terms, and thus smaller overflow (Table 7.18).

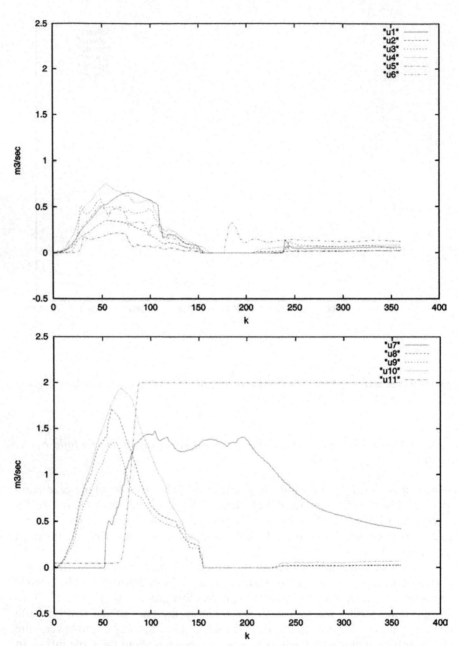

Figure 7.35. Scenario 2: Reservoir outflows $u_i(k)$ for multivariable regulator without feedforward terms; $T = 60$ seconds.

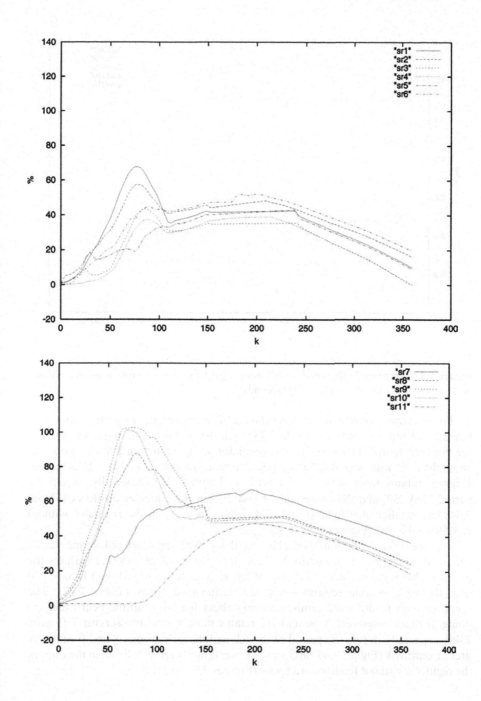

Figure 7.36. Scenario 2: Relative reservoir storages $(V_i(k)/V_{i,max})100\%$ for multivariable regulator without feedforward terms; $T = 60$ seconds.

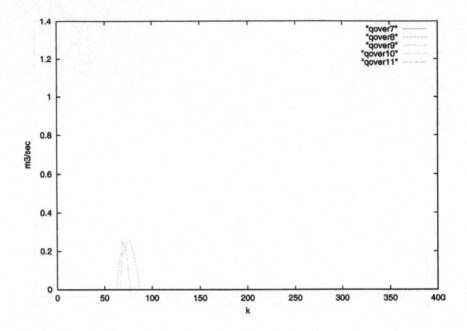

Figure 7.37. Scenario 2: Reservoir overflows $q_{over,i}(k)$ for multivariable regulator without feedforward terms for i=7,...11; T = 60 seconds.

For scenario 2 of external inflows the multivariable regulator with feedforward terms gives fewer overflows (Table 7.18) than the multivariable regulator without feedforward terms (Table 7.17). The regulator with feedforward terms, knowing about the large inflow peaks that are going to reach reservoirs 8, 9 and 10 at around 1 hours, retains more water in reservoir 7 (Figure 7.45), especially during the period 53≤k≤59, and thus reservoirs 9 and 10 can have greater outflows (Figure 7.44) and smaller overflows (Figure 7.46) than the ones in the regulator without feedforward terms (Figures 7.35 to 7.37).

For scenario 3 the multivariable regulator with feedforward terms has a behaviour analogous to scenario 2. Thus, for scenario 3 of external inflows the multivariable regulator with feedforward terms gives fewer overflows (Table 7.18) than the multivariable regulator without feedforward terms (Table 7.17). The regulator with feedforward terms, knowing about the large inflow peaks that are going to reach reservoirs 8, 9, and 10, retains more water in reservoir 7 (Figure 7.48), especially during the period 58≤k≤62, and thus, reservoirs 8 and 9 can have greater outflows (Figure 7.47) and smaller overflows (Figure 7.49) than the ones in the regulator without feedforward terms (Figures 7.38 to 7.40).

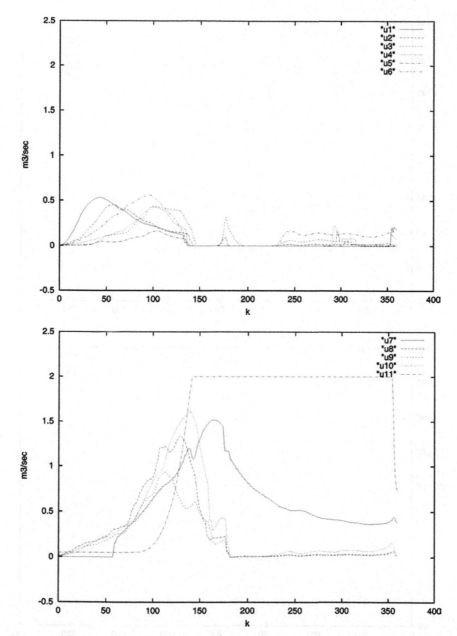

Figure 7.38. Scenario 3: Reservoir outflows $u_i(k)$ for multivariable regulator without feedforward terms; T = 60 seconds.

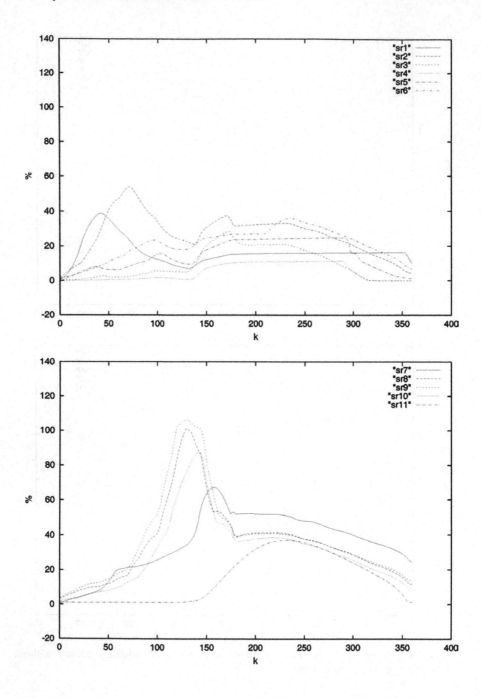

Figure 7.39. Scenario 3: Relative reservoir storages $(V_i(k)/V_{i,max})100\%$ for multivariable regulator without feedforward terms; T = 60 seconds.

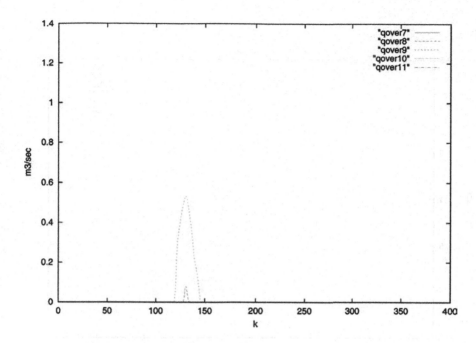

Figure 7.40. Scenario 3: Reservoir overflows $q_{over,i}(k)$ for multivariable regulator without feedforward terms for i=7,...11; T = 60 seconds.

Table 7.18. Reservoir overflows and overload of reservoir 7 in [m^3] for multivariable regulator with feedforward terms.

Reservoir	Scenario 1	Scenario 2	Scenario 3
1	134	0	0
2	218	0	0
3	587	0	0
4	0	0	0
5	220	0	0
6	3	0	0
8	0	0	11
9	0	217	471
10	0	83	0
11	0	0	0
Total	1162	300	482
7	0	0	0

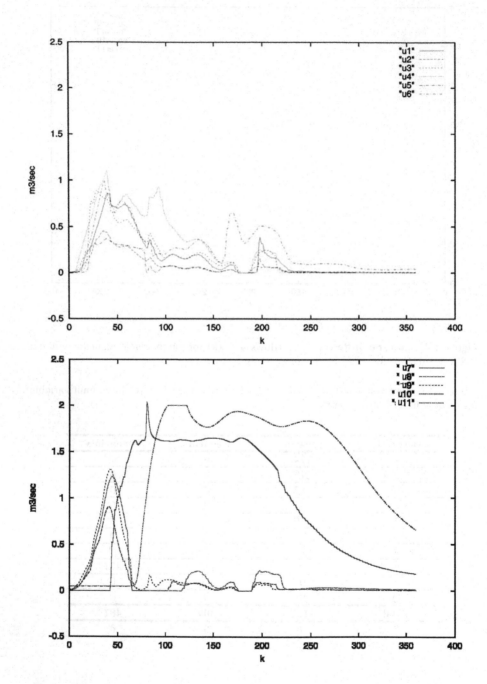

Figure 7.41. Scenario 1: Reservoir outflows $u_i(k)$ for multivariable regulator with feedforward terms; T = 60 seconds.

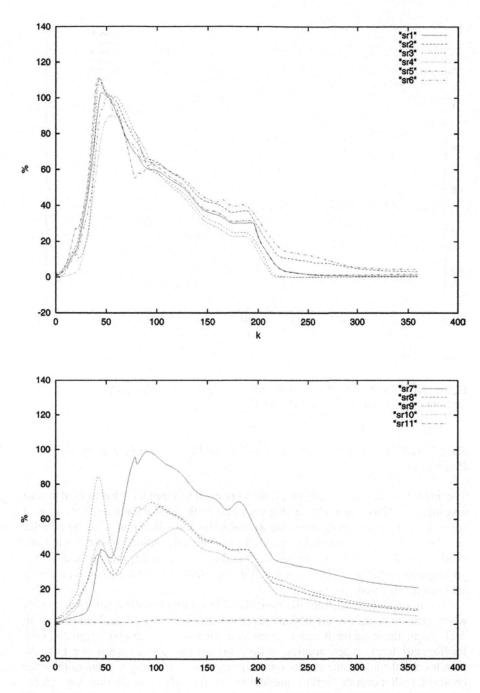

Figure 7.42. Scenario 1: Relative reservoir storages $(V_i(k)/V_{i,max})100\%$ for multivariable regulator with feedforward terms; $T = 60$ seconds.

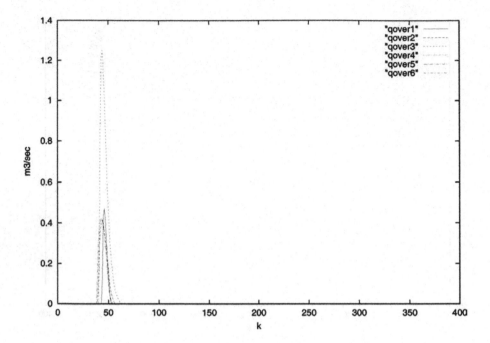

Figure 7.43. Scenario 1: Reservoir overflows $q_{over,i}(k)$ for multivariable regulator with feedforward terms for i=1,...6; T = 60 seconds.

7.3.2.2 Multivariable Regulator with Feedforward Terms and Inaccurate Inflow Predictions

The impact of inaccurate inflow predictions on the regulator's behaviour is also investigated. Thus, the multivariable regulator with feedforward terms is applied when accurate inflow predictions are available for only 30 minutes, or 60 minutes or when there are no available predictions. The calculation of the predictions needed for the prediction horizon K_s (Section 5.4) is made by the use of the prolongation scheme of Section 4.4 and in the third case, past inflow values only are used for the prediction.

The results obtained from the multivariable regulator with feedforward terms when inaccurate inflow predictions are used are summarized in Tables 7.19 to 7.21. From these tables it can be seen that for the multivariable regulator with feedforward terms when accurate inflow predictions are available for only 30 minutes or 60 minuts, the results for scenarios 1 to 3 are very similar to the ones obtained with accurate inflow prediction for the whole prediction horizon K_s.

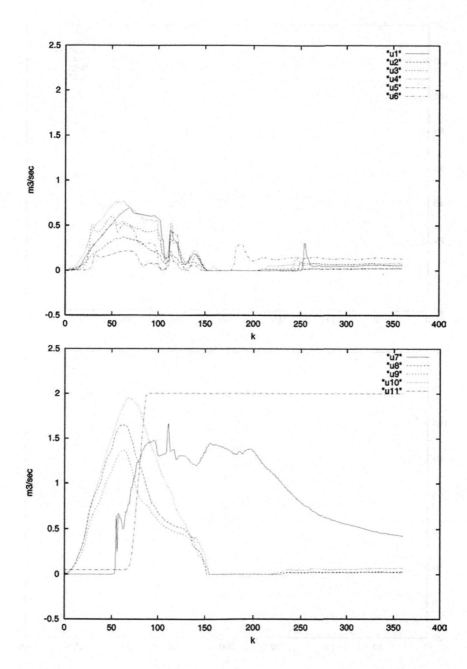

Figure 7.44. Scenario 2: Reservoir outflows $u_i(k)$ for multivariable regulator with feedforward terms; T = 60 seconds.

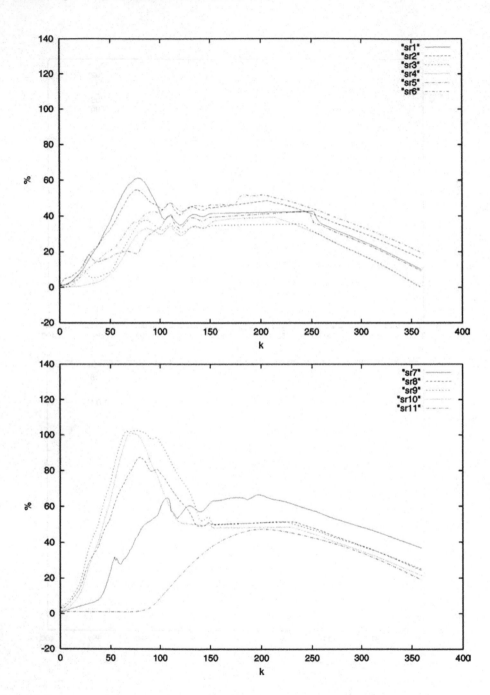

Figure 7.45. Scenario 2: Relative reservoir storages $(V_i(k)/V_{i,max})100\%$ for multivariable regulator with feedforward terms; $T = 60$ seconds.

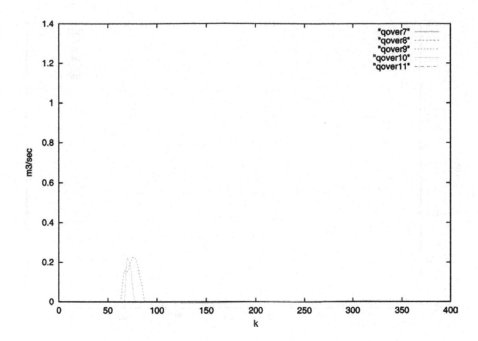

Figure 7.46. Scenario 2: Reservoir overflows $q_{over,i}(k)$ for multivariable regulator with feedforward terms for i=7,…11; T = 60 seconds.

However, when only past values are used for the prediction and, thus, an underestimation or overestimation of the future inflow values is more likely, the results obtained may not always be as good as the ones obtained with accurate inflow predictions. Thus, in the third case, the results for scenario 1 are slightly better than the ones with accurate inflow predictions; for scenario 2 they are slightly inferior to the ones with accurate inflow predictions but are quite similar to the ones of the multivariable regulator without feedforward terms; finally, for scenario 3 the results are similar to the ones of the multivariable regulator with feedforward terms and accurate inflow predictions. These results demonstrate the ability of the multivariable regulator to solve the sewer network control problem even when no accurate inflow predictions are available.

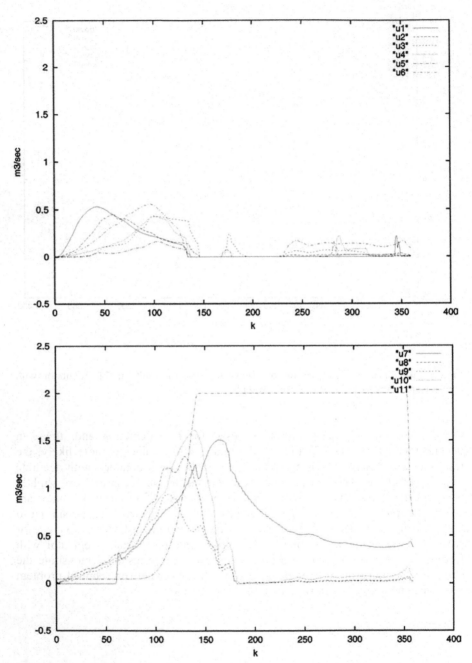

Figure 7.47. Scenario 3: Reservoir outflows $u_i(k)$ for multivariable regulator with feedforward terms; $T = 60$ seconds.

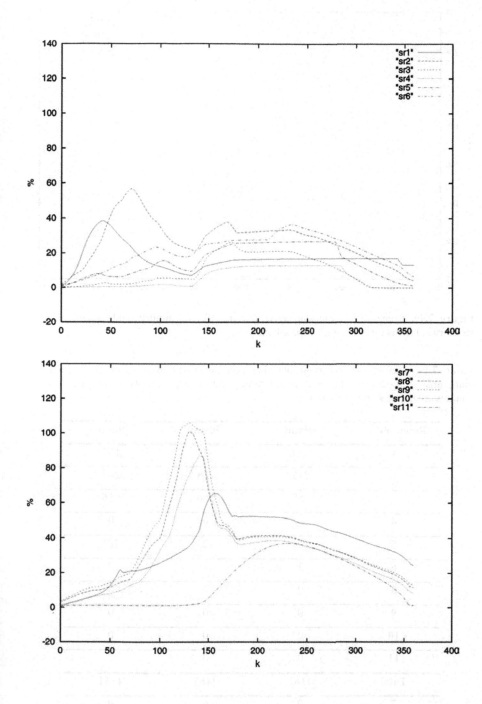

Figure 7.48. Scenario 3: Relative reservoir storages $(V_i(k)/V_{i,max})100\%$ for multivariable regulator with feedforward terms; $T = 60$ seconds.

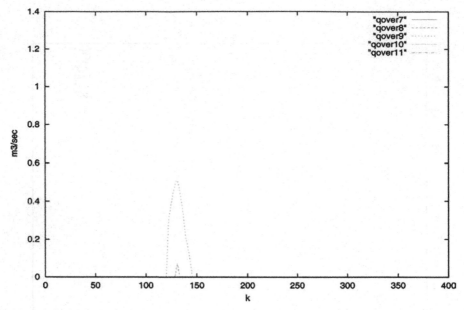

Figure 7.49. Scenario 3: Reservoir overflows $q_{over,i}(k)$ for multivariable regulator with feedforward terms for i=1,...6; T = 60 seconds.

Table 7.19. Scenario 1: Reservoir overflows and overload of reservoir 7 in [m^3] for multivariable regulator with feedforward terms and accurate inflow predictions for 60 minutes, 30 minutes and 0 minutes.

Reservoir	60min	30min	0min
1	134	134	103
2	218	218	204
3	587	586	616
4	0	0	0
5	220	222	221
6	3	3	7
8	0	0	0
9	0	0	0
10	0	0	0
11	0	0	0
Total	1162	1163	1151
7	**0**	**0**	**0**

Table 7.20. Scenario 2: Reservoir overflows and overload of reservoir 7 in [m^3] for multivariable regulator with feedforward terms and accurate inflow predictions for 60 minutes, 30 minutes and 0 minutes.

Reservoir	60 min	30 min	0 min
1	0	0	0
2	0	0	0
3	0	0	0
4	0	0	0
5	0	0	0
6	0	0	0
8	0	0	0
9	217	218	246
10	83	83	98
11	0	0	0
Total	300	301	344
7	**0**	**0**	**0**

Table 7.21. Scenario 3: Reservoir overflows and overload of reservoir 7 in [m^3] for multivariable regulator with feedforward terms and accurate inflow predictions for 60 minutes, 30 minutes and 0 minutes.

Reservoir	60 min	30 min	0 min
1	0	0	0
2	0	0	0
3	0	0	0
4	0	0	0
5	0	0	0
6	0	0	0
8	11	11	11
9	471	472	473
10	0	0	0
11	0	0	0
Total	482	483	484
7	**0**	**0**	**0**

7.4 Comparison Between Nonlinear Optimal Control and Multivariable Feedback Control

The relative efficiency of the nonlinear optimal control using a closed-loop control structure (optimization with rolling horizon) and the multivariable feedback control with and without feedforward terms for the central sewer network control problem is now compared based on their respective control results.

For *scenario 1* of external inflows the rolling horizon with complete inflow information   (Section 7.2.3.2) gives, for all combinations of the repetition periods k_R and the optimization horizons K, better control results than the ones of the multivariable regulator without feedforward terms (Section 7.3.1) and of the multivariable regulator with feedforward terms with accurate (Section 7.3.2.1) or inaccurate inflow predictions (Section 7.3.2.2). When the rolling horizon with incomplete inflow information (Section 7.2.3.3) is used and inflow predictions are only available for 30 minutes or 60 minutes, the control results are slightly superior to the ones of both regulators. However, when inflow predictions are not available at all, the optimization results are similar or slightly inferior (for some combinations of the repetition period k_R and the optimization horizon K) to the ones of both regulators.

For *scenario 2* of external inflows the rolling horizon with complete inflow information (Section 7.2.3.2) gives, in almost all cases, better results than the multivariable regulator without feedforward terms (Section 7.3.1) and the multivariable regulator with feedforward terms with accurate (Section 7.3.2.1) or inaccurate inflow predictions (Section 7.3.2.2). The rolling horizon with incomplete inflow information (Section 7.2.3.3) gives for K_p=20 and K_p=10 equal or slightly inferior control results, whereas for K_p=0 it gives clearly inferior control results compared to the ones of the multivariable regulator without feedforward terms (Section 7.3.1) and of the multivariable regulator with feedforward terms with accurate (Section 7.3.2.1) or inaccurate inflow predictions (Section 7.3.2.2). This is mainly due to the fact that both regulators give larger outflow values for some links (mainly for links 2 and 4) than the ones of the rolling horizon (this can affect the performance of nonlinear optimal control as explained in Section 6.1).

For *scenario 3* of external inflows the rolling horizon with complete inflow information (Section 7.2.3.2) gives for all combinations of the repetition periods k_R and the optimization horizons K superior control results than the multivariable regulator without feedforward terms (Section 7.3.1) and the multivariable regulator with feedforward terms with accurate inflow predictions (Section 7.3.2.1) or inaccurate inflow predictions (Section 7.3.2.2). When the rolling horizon with incomplete inflow information (Section 7.2.3.3) is used and inflow predictions are available for 60 minutes or 30 minutes, the control results are superior to the ones of both regulators, but when inflow predictions are not available, the optimization results are quite inferior to the ones of both regulators. The reason for this latter case is similar to the one of scenario 2, but in this scenario more links have higher outflows with the multivariable regulators (links 2, 4, 5, and 6).

These results indicate that optimization with rolling horizon and multivariable regulators with and without feedforward terms deliver generally a similar quality of control results for the particular sewer network control problem and the investigated inflow scenarios [although, the lack of inflow prediction leads, in some cases, to a relative deterioration of the optimal control approach, particularly if the link flow constraints in the simplified model are selected in a conservative way]. Thus, taking into account that the regulator needs much lower on-line computational effort (Section 5.5) than nonlinear optimal control (Section 4.3.7) and the simplicity of the regulator's computer code, the multivariable regulator may be considered as a valid alternative to the nonlinear optimal control, at least on the basis of the available evidence.

7.5 Concluding Remarks

The results obtained in the previous sections for the particular sewer network control problem demonstrate the efficiency of the nonlinear optimal control when applied in a closed-loop manner and of the multivariable regulator with and without feedforward terms. The nonlinear optimal control approach gave efficient control results for the closed-loop control structure whereas for the open-loop control structure the results were not satisfactory with respect to the overload of reservoir 7. Both regulators, with and without feedforward terms, gave satisfactory control results. A comparison between both approaches, the nonlinear optimal control and the multivariable feedback control, shows that both are applicable with high benefit to the sewer network control problem. Especially, the results of all approaches, when compared to the no-control case, demonstrate the need and efficiency of a central control system for sewer networks.

Chapter 8
Conclusions and Future Research

In this monograph, a generic problem formulation for the central sewer network control has been presented. For the study of the sewer network control problem, two mathematical models, a *realistic simulation model* (accurate model of the sewer network) and a simpler *control design model* (simplified model of the sewer network) were used and described in detail. For the development of the accurate model of the sewer network all the processes in the different elements of the sewer networks were modelled using known laws of hydraulics, whereas in the simpler control design model several simplifications were introduced so as to keep the computational and design effort for control within reasonable levels. In addition, the simulation program KANSIM based on the accurate model of the sewer network was presented.

Two methods for the central control of combined sewer networks, namely a *nonlinear optimal control* and a *multivariable feedback control*, were developed and analysed. Several improvements, modifications, and extensions were introduced to previously developed versions of these methods in order to increase their efficiency and applicability range.

In nonlinear optimal control the main control objectives and the secondary operational objectives of sewer network control are considered directly, via formulation of a nonlinear cost function that is minimized taking into account the state equation and the constraints. A *feasible direction algorithm*, based on the discrete maximum principle, has been used for the solution of the nonlinear optimal control problem. In its present form, the algorithm has been extended to consider directly the state-dependent control constraints, which improves significantly the computational efficiency of the algorithm. The optimal control problem of the sewer network was embedded in a real-time closed-loop *rolling horizon* procedure with repeated optimization runs.

For the development of the linear multivariable feedback regulator the *linear-quadratic methodology* has been used. Application of the linear-quadratic design procedure requires a number of problem simplifications, such as model linearization, quadratic criterion, and no constraints, and includes precise specifications on model structure, model equations, nominal steady-state choice, and quadratic criterion choice. Using this method, multivariable regulators with and without feedforward terms were developed.

To assess the efficiency of both methodologies in satisfying the control objectives of a sewer network control system (whose main task is the minimization of overflows for any rainfall event), when applied to a real sewer network, an extended investigation was performed for the sewer network of Obere Iller (in Bavaria, Germany). This network connects five neighboring cities to one treatment plant. Three scenarios of external inflows were used to investigate the efficacy of the multivariable control law and the nonlinear optimal control for the particular network under a variety of circumstances, and the behaviour of the control methods was investigated assuming availability of both accurate and inaccurate inflow predictions. The simulation program KANSIM was used as a basis for testing and comparing the control performance of both control methods. This program was also used to simulate the no-control case, so as to illustrate the achievable improvements via application of efficient central control strategies to the particular network.

The nonlinear optimal control approach (based on the simplified model of the sewer network) when applied to the particular sewer network control problem assuming availability of accurate inflow predictions provided *automatically* highly efficient flow control within few minutes. The calculated optimal state and control trajectories demonstrate the efficiency of the optimal control approach to address the central sewer network control problem. When the nonlinear optimal control was applied in an open-loop manner, the results were not satisfactory. This was due to the simplifications of the model used in the nonlinear optimal control, due to the parameters of this model of the sewer network that were empirically estimated, and finally due to the fact that in the open-loop control structure no process measurements were utilized. When the nonlinear optimal control was tested using a closed-loop control structure, the results obtained were very satisfactory. For all the investigated scenarios and for all the combinations of the repetition period k_R and the optimization horizon K; the results were very efficient and significantly better than the ones for the no-control case. When the rolling horizon optimal control with incomplete inflow information was used, the control results obtained were similar or slightly inferior to the ones obtained with the rolling horizon with complete inflow information but still much better than the ones of the no-control case.

For the multivariable regulator without feedforward terms, the obtained control and state trajectories demonstrate its efficiency to solve the central sewer network control problem. The results obtained for the investigated scenarios were very satisfactory and were significantly better than the ones obtained when no control actions were taken. The results obtained using the multivariable regulator with feedforward terms, when accurate inflow predictions were available, were equally efficient or slightly superior to the control results obtained using the multivariable regulator without feedforward terms, depending on the particular inflow event. The results obtained from the multivariable regulator with feedforward terms, when inaccurate inflow predictions were used, were very similar to the ones obtained with accurate inflow prediction for the whole prediction horizon.

A comparison between both control methods indicated that optimization with rolling horizon and multivariable regulators with and without feedforward terms deliver a similar quality of control results (with optimisation with rolling horizon giving better results especially when complete inflow information is available) for the particular sewer network control problem and the investigated scenarios. Thus, both methods can be regarded as very efficient methods for the solution of the sewer network control problem with the multivariable regulator having the advantages of much lower on-line computational effort and a much simpler computer code than the nonlinear optimal control. It should also be noted that both control methods are robust with respect to noise in measurements.

Future research is intended to be focused on the global sewer system approach. More specifically an integrated approach that takes into account all parts of the sewer system (sewers, storage elements, treatment plants, receiving waters) will be considered. The combination of the control technologies, the sewer network flow control (applying the methods developed in this monograph), and the treatment technologies (which are being used in order to achieve pollutant removal to meet water quality goals) will be investigated in order to obtain a more efficient solution than the one obtained when sewer network flow control or treatment technologies are applied in an independent way.

References

Anderson BDO, Moore JB (1990) Optimal Control-Linear Quadratic Methods. Prentice Hall, Englewood Cliffs New Jersey

Ashley RM, Hvitved-Jacobsen T, Bertrand-Krajewski JL (1999) Quo Vadis Sewer Process Modelling? Wat Sci Tech 39(9):9-22

Bell W, Johnson G, Winn C B (1973) Simulation and Control of Flow in Combined Sewers. 6th Annual Simulation Symposium Tampa 26-47.

Béron P, Richard D (1982) Simplified Routing in Combined or Storm Interceptors. Urban Drainage Systems, Proc of the 1st Int Seminar, Featherstone RE, James A (eds.) Southampton 2-67-76

Bradford B (1977) Optimal Storage Control in a Combined Sewer System. Journal of the Water Resources Planning and Management Division. Proc ASCE 103:1-15

Bretschneider H, Lecher K, Schmidt M (1982) Taschenbuch der Wasserwirtschaft. Verlag, Paul Parey, Hamburg

Bunday BD (1984) Basic Optimisation Methods. Edward Arnold Publishers, London

Chebbo G, Gromaire MC, Ahyerre M, Garnaud S (2001) Production and Transport of Urban Wet Weather Pollution in Combined Sewer Systems: the "Marais" Experimental Urban Catchment in Paris. Urban Water 3:3-15

Chu WS, Yeh WWG (1978) A Nonlinear Programming Algorithm for Real-Time Hourly Reservoir Operations. Water Resources Bulletin, American Water Resources Association 14(5):1048-1063

Dahlin EB, Shen DWC (1966) Optimal Solution to the Hydro-Steam Dispatch Problem for Certain Practical Systems. IEEE Transactions on Power Apparatus and Systems 85(5):437-454

Dorato P, Abdallah C, Cerone V (1995) Linear-Quadratic Control: An Introduction. Prentice Hall, Englewood Cliffs, New Jersey

Duchesne S, Mailhot A, Dequidt E, Villeneuve JP (2001) Mathematical Modeling of Sewers Under Surcharge for Real Time Control of Combined Sewer Overflows. Urban Water 3:241-252

Duncan WJ, Thom AS, Young AD (1970) Mechanics of Fluids. 2nd ed. Pitman Press, Bath

Erbe V, Frehmann T, Geiger WF, et al (2002) Integrated Modelling as an Analytical and Optimisation Tool for Urban Watershed Management. Wat Sci Tech 46(6-7):141-150

Fletcher R (1987) Practical Methods of Optimization. 2nd ed. John Wiley & Sons, Chichester

Foufoula-Georgiou E, Kitanidis PK (1988) Gradient Dynamic Programming for Stochastic Optimal Control of Multidimensional Water Resources Systems. Wat Res Res 24(8):1345-1359

Frehmann T, Nafo I, Niemann A, Geiger WF (2002) Storm Water Management in an Urban Catchment: Effects of Source Control and Real-Time Management of Sewer Systems on Receiving Water Quality. Wat Sci Tech 46(6-7):19-26

Froise S, Burges SJ (1978) Least-Cost Design of Urban-Drainage Networks. Journal of the Water Resources Planning and Management Division, Proc ASCE 104:75-92

Fuchs L, Beeneken T, Spönemann P, Scheffer C (1997) Model Based Real-Time Control of Sewer System Using Fuzzy-Logic. Wat Sci Tech 36(8-9):343-347

Garcia A, Hubbard M, De Vries JJ (1992) Open Channel Transient Flow Control by Discrete Time LQR Methods. Automatica 28(2):255-264

Gelormino MS, Ricker NL (1994) Model-Predictive Control of a Combined Sewer System. Int J Control 59(3): 793-816

Gill PE, Murray W, Wright MH (1981) Practical Optimization. Academic Press, London

Grygier JC, Stedinger JR (1985) Algorithms for Optimizing Hydropower System Operations. Wat Res Res 21(1):1-10

Gutman PO (1986) A Linear Programming Regulator Applied to Hydroelectric Reservoir Level Control. Automatica 22(5):533-541

Hano I, Tamura Y, Narita S (1966) An application of the Maximum Principle to the Most Economical Operation of Power Systems. IEEE Transactions on Power Apparatus and Systems 85(5):486-494

Heidari M, Chow VT, Kokotović PV, Meredith DD (1971) Discrete Differential Dynamic Programming Approach to Water Resources Systems Optimization. Wat Res Res 7(2):273-282

Hernebring C, Jönsson LE, Thorén UB, Møller A (2002) Dynamic Online Sewer Modelling in Helsingborg. Wat Sci Tech 45(4-5):429-436

Jacopin C, Lucas E, Desbordes M, Bourgogne P (2001) Optimisation of Operational Management Practices for the Detention Basins. Wat Sci Tech 44(2-3):277-285

Ji Z, Vitasovic Z, Zhou S (1996) A Fast Hydraulic Numerical Model for Large Sewer Collection Systems. Wat Sci Tech 34(3-4):17-24

Katebi R, Johnson MA, Wilkie J (1999) Control and Instrumentation for Wastewater Treatment Plants. Advances in Industrial Control, Springer-Verlag, London

Kirk DE (1970) Optimal Control Theory: An Introduction. Prentice Hall, Englewood Cliffs, New Jersey

Klepiszewski K, Schmitt TG (2002) Comparison of Conventional Rule Based Flow Control with Control Processes Based on Fuzzy Logic in a Combined Sewer System. Wat Sci Tech 46(6-7):77-84

Labadie JW, Morrow DM, Chen YH (1980) Optimal Control of Unsteady Combined Sewer Flow. Journal of the Water Resources Planning and Management Division, Proc ASCE 106:205-223

Lambert JD (1991) Numerical Methods for Ordinary Differential Systems. The Initial Value Problem. John Wiley & Sons, Chichester

Lee JH, Bang, KW (2000) Characterization of Urban Stormwater Runoff. Wat Res 34(6):1773-1780

Lee YM, Ellis, JH (1996) Comparison of Algorithms for Nonlinear Integer Optimization: Application to Monitoring Network Design. J Env Eng 122(6):524-531

Liu Y, Wu W (1993) The Modelling and Control of Large Scale Water Distribution Systems. 12th World Congress International Federation on Automatic Control, Preprints of Papers, Sydney 9:527-530

Marinaki M (1995) Central Flow Control in Sewer Networks. M.Sc. Thesis. Technical University of Crete, Chania, Greece (in Greek)

Marinaki M (2002) Optimal Real-Time Control of Sewer Networks. Ph.D. Thesis. Technical University of Crete, Chania, Greece

Marinaki M, Papageorgiou M (1995) Optimal Control of Sewer Network Flow. Internal Report No. 1995-5, Dynamic Systems and Simulation Laboratory, Technical University of Crete, Chania, Greece

Marinaki M, Papageorgiou M (1996a) A Multivariable Regulator Approach to Central Sewer Network Flow Control. Internal Report No. 1996-7, Dynamic Systems and Simulation Laboratory, Technical University of Crete, Chania, Greece

Marinaki M, Papageorgiou M (1996b) A LQ-Regulator with Feedforward Terms Applied to Sewer Network Flow Control. 4th International Conference on Control, Automation, Robotics and Vision (ICARCV'96), Singapore, December 3-6, pp 1441-1445

Marinaki M, Papageorgiou M (1997a) Central Flow Control in Sewer Networks. ASCE Journal of Water Resources Planning and Management 123(5):274-283

Marinaki M, Papageorgiou M (1997b) Central Flow Control in Sewer Networks. Proc of the 7th Panhellenic Conference of the Greek Hydrotechnical Association, Patra, Greece, October 14-18, pp 493-502 (in Greek)

Marinaki M, Papageorgiou M (1998) Nonlinear Optimal Flow Control for Sewer Networks. 1998 American Control Conference, Philadelphia, Pennsylvania, June 24-26, pp 1289-1293

Marinaki M, Papageorgiou M (1999) A Non-linear Optimal Control Approach to Central Sewer Network Flow Control. International Journal of Control, 72(5):418-429

Marinaki M, Papageorgiou M (2001) Rolling-Horizon Optimal Control of Sewer Networks. Proc of the 2001 IEEE International Conference on Control Applications, México City, México, September 5-7, pp 594-599

Marinaki M, Papageorgiou M, Messmer A (1999) Multivariable Regulator Approach to Sewer Network Flow Control. ASCE Journal of Environmental Engineering 125(3):267-276

Masse B, Zug M, Tabuchi JP, Tisserand B (2001) Long Term Pollution Simulation in Combined Sewer Networks. Wat Sci Tech 43(7):83-89

Mays LW, Tung YK (1978) State Variable Model for Sewer Network Flow Routing. Journal of the Environmental Engineering Division, Proc ASCE 104:15-30

McLaughin D, Velasco HL (1990) Real-Time Control of a System of Large Hydropower Reservoirs. Wat Res Res 26(4):623-635

Meirlaen J, Van Assel J, Vanrolleghem PA (2002) Real Time Control of the Integrated Urban Wastewater System Using Simultaneously Simulating Surrogate Models. Wat Sci Tech 45(3):109-116

Meredith DD (1975) Optimal Operation of Multiple Reservoir System. J Hyd Div 101:299-312

Messmer A (1998) KANSIM Documentation. IB Messmer, Seeshaupt, Germany

Messmer A, Papageorgiou M (1992) Multireservoir Sewer-Network Control via Multivariable Feedback. Journal of Water Resources Planning and Management 118(6):585-602

Méthot JF, Pleau M (1997) The Effects of Uncertainties on the Control Performance of Sewer Networks. Wat Sci Tech 36(5):309-315

Mizyed NR, Loftis JC, Fontane DG (1992) Operation on Large Multireservoir Systems Using Optimal Control Theory. Journal of Water Resources Planning and Management 118(4):371-387

Murray DM, Yakowitz SJ (1979) Constrained Differential Dynamic Programming and its Application to Multireservoir Control. Wat Res Res 15(5):1017-1027.

Nelen F (1994) A Model to Assess the Performance of Controlled Urban Drainage Systems. Wat Sci Tech 29(1-2):437-444

Nielsen CS, Ravn H (1985) Investigation of a New Numerical Method for Control of a Water-Supply Network. IFAC Conference on Systems Analysis Applied to Water and Related Land Resources, Lisbon, Portugal, IV-23-IV-28.

Papageorgiou M (1983) Automatic Control Strategies for Combined Sewer Systems. J of Env Eng 109(6):1385-1402

Papageorgiou M (1985) Optimal Multireservoir Network Control by the Discrete Maximum Principle. Wat Res Res 21(12):1824-1830

Papageorgiou M (1988) Certainty Equivalent Open-Loop Feedback Control Applied to Multireservoir Networks. IEEE Transactions on Automatic Control 33(4):392-399

Papageorgiou M (1996) Optimierung. 2nd edn. Oldenbourg Verlag, Munich, Germany

Papageorgiou M (1997) Automatic Control Methods in Traffic and Transportation. Internal Report No. 1997-10, Dynamic Systems and Simulation Laboratory, Technical University of Crete, Chania, Greece

Papageorgiou M, Marinaki M (1995) A Feasible Direction Algorithm for the Numerical Solution of Optimal Control Problems. Internal Report No. 1995-4, Dynamic Systems and Simulation Laboratory, Technical University of Crete, Chania, Greece

Papageorgiou M, Mayr R (1985) Optimal Real-Time Control of Combined Sewer Networks. IFAC Conference on Systems Analysis Applied to Water and Related Land Resources, Lisbon, Portugal, VIII-17-VIII-22.

Papageorgiou M, Mayr R (1988) Comparison of Direct Optimization Algorithms for Dynamic Network Flow Control. Optimal Control Applications & Methods 9:175-185

Papageorgiou M, Messmer A (1985) Continuous-Time and Discrete-Time Design of Water Flow and Water Level Regulators. Automatica 21(6):649-661

Papageorgiou M, Messmer A (1989) Flow Control of a Long River Stretch. Automatica 25(2):177-183

Papageorgiou M, Mevius F (1985) Hierarchical Control Strategy for Combined Sewer Networks. Instrumentation and Control of Water and Wastewater Treatment and Transport Systems, R. A. R. Drake, ed., Pergamon Press, Oxford, U.K., pp 209-216

Papanikas DG (1981) Applied Fluid Mechanics. Vol. I. 2nd ed. ACHAIOS, Patras, Greece (in Greek)

Pleau M, Méthot F, Lebrun AM, Colas H (1996) Minimizing Combined Sewer Overflows in Real-Time Control Applications. Wat Qual Res J Canada 31(4):775-786

Pleau M, Pelletier G, Colas H, Lavallée P, Bonin R (2001) Global Predictive Real-Time Control of Quebec Urban Community's Westerly Sewer Network. Wat Sci Tech 43(7): 123-130

Rauch W, Bertrand-Krajewski JL, Krebs P, et al (2002) Deterministic Modelling of Integrated Urban Drainage Systems. Wat Sci Tech 45(3):81-94

Riedmiller M (1994) Advanced Supervised Learning in Multi-layer Perceptrons – From Backpropagation to Adaptive Learning Algorithms. Computer Standards and Interfaces 16(3): 265-278

Riedmiller M, Braun H (1993) A Direct Adaptive Method for Faster Backpropagation Learning: The RPROP Algorithm. Proc IEEE International Conference on Neural Networks, San Francisco, CA

Robinson DK, Labadie JW (1981) Optimal Design of Urban Storm Water Drainage Systems. 1981 International Symposium on Urban Hydrology, Hydraulics, and Sediment Control, Lexington, Kentucky, 145-156

Rosen JB (1960) The Gradient Projection Method for Nonlinear Programming. Part I. Linear Constraints. Journal of Society of Industrial and Applied Mathematics 8(1):181- 217

Saha TN, Khaparde SA (1978) An Application of a Direct Method to the Optimal Scheduling of Hydrothermal System. IEEE Transactions on Power Apparatus and Systems 97(3):977-983

Sakr AF, Dorrah, HT (1985) Optimal Control Algorithm for Hydropower Plants Chain Short-Term Operation. Preprints of 7th Conference on Digital Computer Applications to Process Control, IFAC, Vienna, Austria, pp 171-177

Scales LE (1985) Introduction to Non-linear Optimization. Macmillan Publishers, London

Schütze MR, Butler D, Beck MB (2002) Modelling, Simulation and Control of Urban Wastewater Systems. Springer-Verlag, London

Seidl M, Servais P, Mouchel JM (1998) Organic Matter Transport and Degradation in the River Seine (France) after a Combined Sewer Overflow. Water Resources 32(12):3569-3580

Sirisena HR, Halliburton TS (1981) Long-Term Optimization of Hydro-Thermal Power Systems by Generalized Conjugate-Gradient Methods. Optimal Control Applications & Methods 2:351-364

Walters GA (1985) The Design of the Optimal Layout for a Sewer Network. Engineering Optimization 9:37-50

Wanka K, Königer W (1984) Unsteady Flow Simulation in Complex Drainage Systems by HVM-Hydrograph Volume Method. Proc 1st International Conference on Channels and Channel Control Structures, Southampton, England, 5.17-31.

Wardlaw R, Sharif M (1999) Evaluation of Genetic Algorithms for Optimal Reservoir System Operation. Journal of Water Resources Planning and Management 125(1):25-33

Winn CB, Moore JB (1973) The Application of Optimal Linear Regulator Theory to a Problem in Water Pollution. IEEE Transactions on Systems, Man, and Cybernetics 3(5):450-455

Yeh WWG, Becker L (1982) Multiobjective Analysis of Multireservoir Operations. Wat Res Res 18(5):1326-1336

Zabel T, Milne I, Mckay G (2001) Approaches Adopted by the European Union and Selected Member States for the Control of Urban Pollution. Urban Water 3:25-32

Zessler U, Shamir U (1989) Optimal Operation of Water Distribution Systems. Journal of Water Resources Planning and Management 115(6):735-752

Zug M, Faure D, De Belly B, Phan L (2001) Use of Real Time Control Modelling on the Urban Sewage System of Nancy. Wat Sci Tech 44(2-3):261-268

Author Profiles

Magdalene Marinaki was born in Chania, Greece. In 1993 she received the Dipl.-Eng. degree in production engineering and management from the Technical University of Crete, Greece, and she received the M.Sc. and Ph.D. degrees in production engineering and management from the same university in 1995 and 2002, respectively.

She received a scholarship from the Institute of National Scholarships and the Technical Chamber of Greece in the academic year 1991–1992 for high-level performance in her studies. Since March 1994, she has been a research and teaching associate of the Dynamic Systems and Simulation Laboratory of the Technical University of Crete. She has participated in research projects and she has assisted undergraduate courses. Since September 2000, she has been teaching at the Technological Educational Institute of Crete, Branch of Chania. Since September 2002, she has been a contract lecturer at the Technical University of Crete, Department of Production Engineering and Management. She is the author of research reports and papers in international journals and scientific conferences. Her research interests include optimal and automatic control, operations research, and applications to water systems, transportation systems, and further areas. She is a member of the Technical Chamber of Greece (TEE). E-mail address: marinakis@ergasya.tuc.gr.

Markos Papageorgiou was born in Thessaloniki, Greece, in 1953. He received the Diplom-Ingenieur and Doktor-Ingenieur (honors) degrees in electrical engineering from the Technical University of Munich, Germany, in 1976 and 1981, respectively. From 1976 to 1982 he was a research and teaching assistant at the Control Engineering Chair, Technical University of Munich. He was a free associate with Dorsch Consult, Munich (1982–1988), and with Institute National de Recherche sur les Transports et leur Sécurité (INRETS), Arcueil, France (1986–1988). From 1988 to 1994 he was a professor of automation at the Technical University of Munich. Since 1994 he has been a professor at the Technical University of Crete, Chania, Greece. He was a visiting professor at the Politecnico di Milano, Italy (1982), at the Ecole Nationale des Ponts et Chaussées, Paris (1985–1987), and at MIT, Cambridge (1997, 2000); and a visiting scholar at the University of Minnesota (1991, 1993), University of Southern California (1993), and the University of California, Berkeley (1993, 1997, 2001).

Dr. Papageorgiou is the author of the books *Applications of Automatic Control Concepts to Traffic Flow Modeling and Control* (Springer, 1983) and *Optimierung* (Oldenbourg, 1991, 1996), the editor of the *Concise Encyclopedia of Traffic and Transportation Systems* (Pergamon Press, 1991), and the author or co-author of some 230 technical papers. His research interests include automatic control and optimization theory and applications to traffic and transportation systems, water systems and further areas. He is an associate editor of *Transportation Research-Part C*, of *IEEE Transactions on Intelligent Transportation Systems* and of *IEEE Control System Society, Conference Editorial Board*, and chairman of the IFAC Technical Committee on Transportation Systems. He is a member of the Technical Chamber of Greece (TEE) and a fellow of IEEE. He received a DAAD scholarship (1971–1976), the 1983 Eugen-Hartmann award from the Union of German Engineers (VDI), and a Fulbright Lecturing/Research Award (1997). Email address: markos@dssl.tuc.gr.

Index